Soil Moisture Importance

Edited by Ram Swaroop Meena
and Rahul Datta

Published in London, United Kingdom

IntechOpen

Supporting open minds since 2005

Soil Moisture Importance
http://dx.doi.org/10.5772/intechopen.82898
Edited by Ram Swaroop Meena and Rahul Datta

Contributors
Rajan Bhatt, Ram Swaroop Meena, Aleš Kučera, Aleš Bajer, Pavel Samec, Valerie Vranová, Rahul Datta, Keith Skene, Tomáš Vichta, Ning Ai, Tianxing Wei, Qingke Zhu, Guangquan Liu, Anita Kumawat, Devideen Yadav, Adrijana Filipović, Muthuraman Yuvaraj, Kasiviswanathan Subash Chandra Bose, Elavarasi Prabakaran, Eman Tawfik Hussien, Kala Samadharmam, Ittyamkandath Rashmi

First published in London, United Kingdom, 2021 by IntechOpen
IntechOpen is the global imprint of INTECHOPEN LIMITED, registered in England and Wales, registration number: 11086078, 5 Princes Gate Court, London, SW7 2QJ, United Kingdom
Printed in Croatia

British Library Cataloguing-in-Publication Data
A catalogue record for this book is available from the British Library

Additional hard and PDF copies can be obtained from orders@intechopen.com

Soil Moisture Importance
Edited by Ram Swaroop Meena and Rahul Datta
p. cm.
Print ISBN 978-1-83968-095-3
Online ISBN 978-1-83968-096-0
eBook (PDF) ISBN 978-1-83968-097-7

For EU product safety concerns:
IN TECH d.o.o., Prolaz Marije Krucifikse Kozulić 3, 51000 Rijeka, Croatia, info@intechopen.com or visit our website at intechopen.com.

Meet the editors

Dr. Ram Swaroop Meena is an assistant professor in the Department of Agronomy, Institute of Agricultural Sciences, Banaras Hindu University, Varanasi, India. Dr. Meena is a recipient of the Raman Research Fellowship from the Ministry of Human Resource Development (MHRD), Government of India (GOI). He completed his postdoctoral research on soil carbon sequestration under Prof. Rattan Lal, Director, Carbon Management and Sequestration Center (CMASC), Columbus, USA, World Food Prize Laureate 2020. Dr. Meena has supervised many postgraduate students and Ph.D. candidates and has more than 10 years of research and teaching experience. He is currently working on three externally funded projects: DST, MHRD, and ICAR. Dr. Meena has published more than 110 research and review papers in indexed journals as well as numerous books and book chapters, both nationally and internationally. He has worked as an expert for the National Council of Educational Research and Training (NCERT), MHRD, and GOI. Dr. Meena contributes to many agricultural extension activities, training, meetings, workshops, and so on.

Dr. Rahul Datta is a soil microbiologist and enzymologist. He obtained an MSc and Ph.D. from the Department of Geology and Pedology, Mendel University, Brno, Czech Republic, where he is currently a researcher. Dr. Datta's research focuses on understanding the effect of biogenic and xenobiotic substances on microbial metabolism in the soil. During his career, Dr. Datta has acquired global experience in soil science research by working with renowned scientists. He worked as a visiting scientist with Dr. Richard Dick at Ohio State University, Ohio, USA, and Prof. Paolo Nannipieri at the University of Florence, Italy. Dr. Datta has published numerous research articles and books and has reviewed 200 papers in peer-reviewed journals. He is currently a reviewer for fifty-eight major scientific journals. Dr. Datta is an editorial board member of Open Agriculture and is hosting a special issue of the Journal of Agriculture and Sustainability. He is also a current member of the Soil Science Society of America.

Contents

Preface

The world is facing many environmental, food, nutritional, social, and economic security challenges. Most alarming is the conservation of natural resources, especially soil and water. Meeting food and nutritional security targets requires the implementation of time demand policies that ensure soil and water retention for sustainable productivity. This book examines ways of improving soil moisture management to support a sustainable ecosystem for the future world.

Ram Swaroop Meena
Institute of Agricultural Sciences,
India

Rahul Datta
Mendel University in Brno,
Czech Republic

Chapter 1

Delineation of Soil Moisture Potentials and Moisture Balance Components

Rajan Bhatt and Ram Swaroop Meena

Abstract

Root architecture in soils is directly affecting crop yield potential, through influencing the moisture potential of soil and its balance components, as only transpiration share is useful for them. Soil moisture potential responsible for the soil moisture curves on the basis of differential energy states is quite important. Generally, a soil moisture flow rate is considered for its kinetic energy. Consequently, soil moisture energy state is defined by its equivalent potential energy, which is by virtue of its place in a force field which could assist to improve the water-use efficiency. Irrigation water losses significantly occur under the flood irrigation through evaporation, seepage, and drainage. While the soil moisture potential declines with help of the tensiometer, and significantly save the irrigation water. For evaluating the performance of any resource conservation technologies (RCT) in the region, estimation of the evapotranspiration (ET) is very important to analyze the effect of the RCT. It is also helpful in balancing the nutrient inflows in the plants through roots, which results to the improvement of land and water productivity. Hence, delineation of the soil moisture potentials and moisture balance components is important to improve the land as well as water productivity; it makes the livelihood security better in the water-stressed regions on the globe. This chapter deals with the methodological part of soil moisture potentials and moisture balance components, which is useful for the policymakers, modelers, scientists, students, and teachers engaged in the irrigation experiments under texturally divergent soils.

Keywords: moisture potential, tensiometer, seepage, drainage, unsaturated hydraulic conductivity, irrigation

1. Introduction

The complex nature of the soil pore space and the water held therein makes it difficult to delineate the soil-water interface and moisture advancements in the soil, which is further influenced by soil matrix geometry. Soil moisture is the amount of moisture present in soil pores, which is a must for all important ecological processes and plays a critical and significant role in all the physiological processes. Throughout the globe, water scarcity is an emerging problem that must be worked out for sustaining agricultural growth [1–3]. Different RCTs are recommended for having improved water productivities across the globe [4–6]. The scientists at NASA's Goddard Space Flight Center generate groundwater and soil moisture drought indicators each week. They are based on terrestrial water storage observations derived from GRACE-FO satellite data and integrated with other observations, using a sophisticated numerical model of land surface water and energy processes. The drought indicators describe current wet or dry conditions, expressed as a percentile showing the probability of occurrence for that particular location and time of year, with lower values (warm colors) meaning dryer than normal, and higher values (blues) meaning wetter than normal (**Figure 1**).

Global analysis based on intermediate population growth rate revealed that water scarcity is a global issue and therefore needs to be addressed for mitigating its adverse effects onto the overall land and water productivities of agricultural crops (**Figure 2**).

Further, in India, the net irrigated area increased from the 1960s and is further projected to increase by 2030 (**Figure 3**), which further increased the installed tube wells and further declined the underground water table of the country, which might be beyond the reach of the poor farmers.

Soil water potential must be understood, and its applications must be applied in field conditions. For measuring the soil water potential, the instrument highlighted as tensiometer is used for irrigating the crops, namely, rice, without affecting the overall land as well as water productivity [7]. Tensiometer measured the soil suction, and when soil dries, then the inner water in the tensiometer via porous cup moves out in the soil. Hence, as a result, the potential reading in tensiometer increased, and at predefined levels of potential, irrigation is applied to crops [1, 7]. After irrigation, water moved back into the tensiometer from the irrigated soil, and water level of inner tube moved back to normal, namely, green level. Soil water potential (as controls moisture movements) is the ultimate technique, under unsaturated conditions when only micropores are water filled, while macropores are air filled for improving the declined water-use efficiency without affecting the grain yields more particularly in global water-stressed regions [7, 8]. However, both macro- and micropores are water filled, and conducting it under saturated soil condition seldom

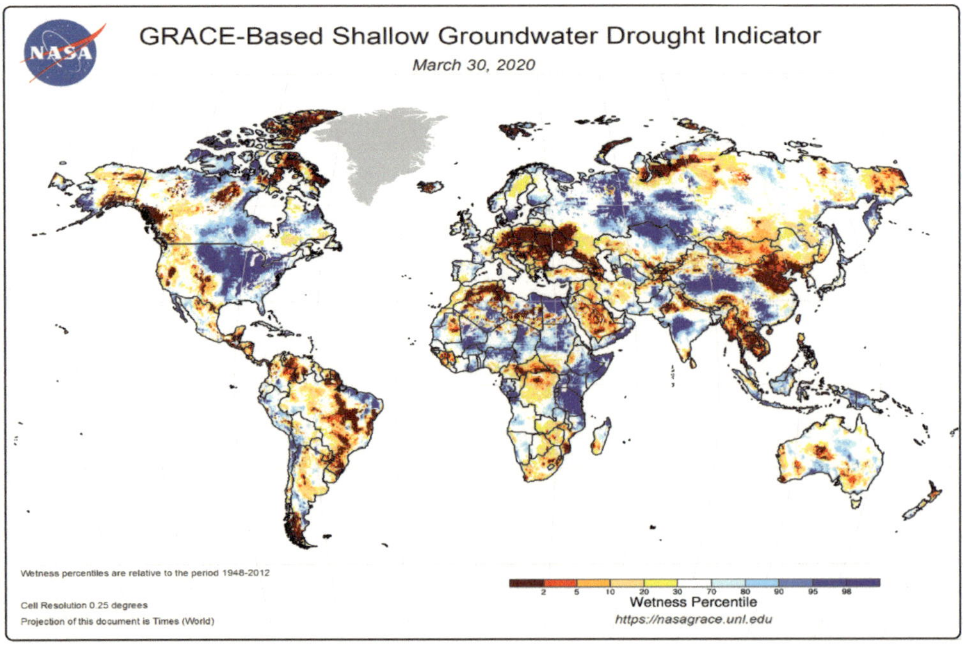

Figure 1.
GRACE based global shallow groundwater drought indicators (https://nasagrace.unl.edu/).

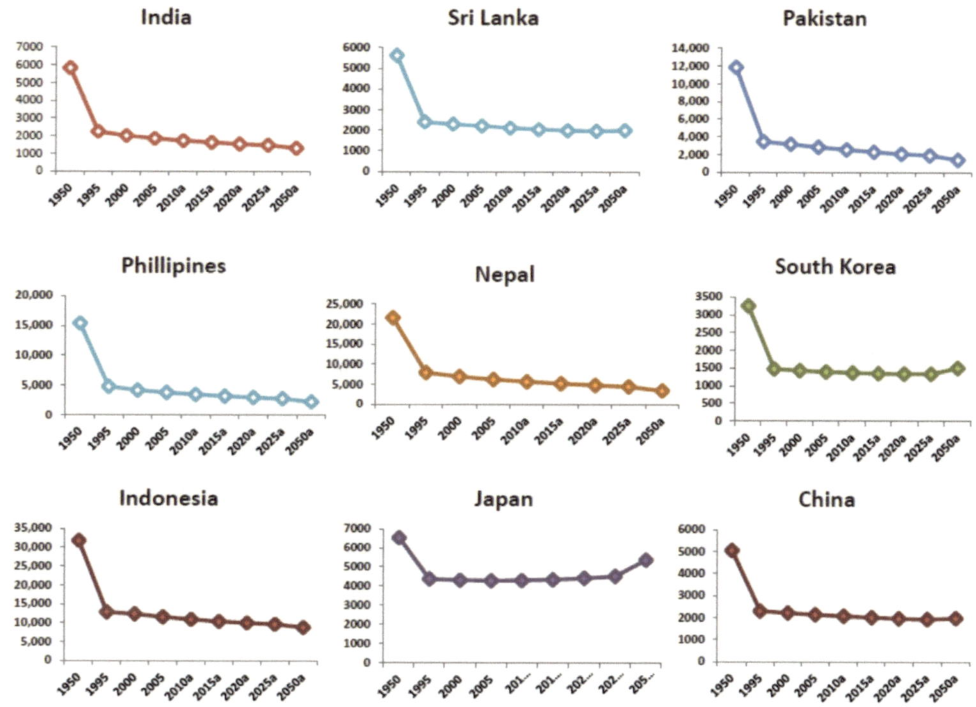

Figure 2.
Water availability Per capita (m³) in chief paddy-growing Asian countries viz-a-viz upcoming years (1950–2050) a Estimate based on the population growth trends Source: Modified from [1]).

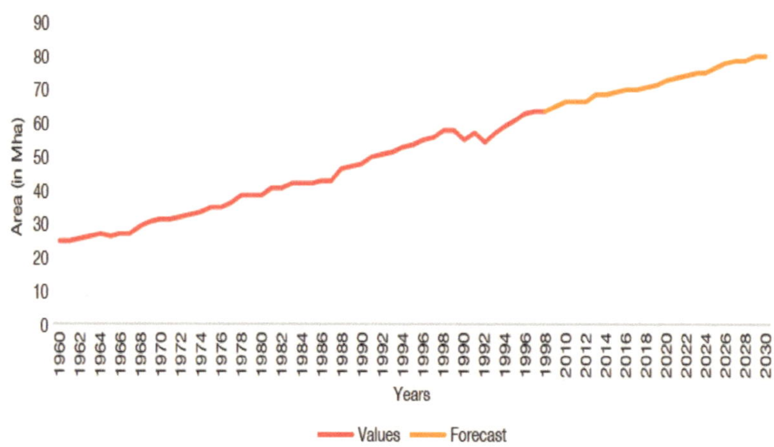

Figure 3.
Net irrigated area in India (Source: Food and Agriculture Organization, 2008).

exists in nature. Gravity and soil water potential are the main driving forces under saturated and unsaturated conditions, responsible for soil moisture movement. Micropores of fine-textured clayey soils are capable of holding water for a longer period of time even at higher value of suction, while macropores of sandy soil drain out the water quickly at a smaller suction. Therefore, generally frequent irrigations resulting in lower water productivity are reported in the sandy soils as compared to the clayey fine-textured soil. In nature, soil moisture has different quantities and forms of energy by virtue of which it moves from one to another point in soil. The potential concept to the soil water in relation to its movement was first given by

Buckingham [9] in his classical paper on the capillary potential, while Gardner [10] showed the dependency of water potential on the water content, and Richards [11] prepared a tensiometer for measuring it. Hence, the concept of soil moisture movement is not new but is still difficult to understand by the new budding students and agricultural scientists dealing with agricultural water management. Moreover, quite often research papers published in reputed journals discussed the water balance components without discussing much on their estimation/calculative part, which further confuses the students. Therefore, estimation of the different soil moisture components is a must so as to perform new water management experiments with clear objectives of having higher water productivity under texturally divergent soils. These RCTs are site and situation specific, and a single RCT is not effective equally in all places for improving the water-use efficiency [12]. Therefore, considering above discussions, this chapter focused on the estimation of components of soil moisture potentials and balance components for the proper understanding of the concept by the end users, namely, agricultural students and even budding scientists, for conduction of more region-specific water management experiments under texturally divergent soils for ultimately improving water productivity without affecting the grain yields in water-stressed regions of the globe.

2. Soil moisture potential (ψW)

Soil moisture potential in the common language is the potential of moisture to do work by its position in soil. ψW is the difference between the activity of the water molecule in pure distilled water and soil solution at normal atmospheric temperature and pressure which might be greater or lesser. In the definition of International Soil Science Society [13], ψW may be defined as "the amount of work that must be done per unit quantity of pure water in order to transport reversibly and isothermally an infinitesimal quantity of water from a pool of pure water at a specified elevation at atmospheric pressure to the soil water (at the point under consideration)." Hence, a reference state is a must.

ψW could also be delineated by knowing in a solution of nonelectrolytes, the chemical potential of water which further depends upon mean free energy per molecule and water molecule concentration. The chemical potential of pure water reduces with the addition of salts, which could be expressed as

$$\psi W = \mu W - \mu W^* = RT \, Ln \, Nw \quad (1)$$

where R is the universal gas constant, T is the absolute temperature, and Nw is the mole fraction of water, respectively.

For the simple ionic solution,

$$\psi W = \mu W - \mu W^* = RT \, Ln \, aW \quad (2)$$

where aw is the activity of the water molecules, which measured how easy the water content may be utilized. Further, *t*he water vapor pressure of the solution expressed as a fraction of the vapor pressure of pure water at the same temperature (or the equilibrium humidity expressed as a fraction) is numerically equal to the activity of the water (a_w) in the solution. Eq. (2) is more useful as water always has ions.

When water contains a number of ions, then

$$\psi W = \mu W - \mu W^* = RT \, Ln \, e/e_o \quad (3)$$

where ψW is the water potential, μW is the solution's water chemical potential, μW* is the pure state's water chemical potential, R is the universal gas constant (82 bars cm-^{2}), T is the absolute temperature, and e/e_o is the relative vapor pressure, respectively.

ψW could be expressed depending upon the units used for the expression of quantity of water.

Expressed units	Units of ψW
Mass	erg g^{-1}
Volume	Dynes cm^2
Weight	cm, m, mm

Among all the units, weight units are more convenient to use.

However, when all pores are water filled, conducting it under saturated conditions, then the actual and potential vapor pressure is the same, and thus e/e_o comes out to be 1 (log 1 = 0). Thus, under saturated soil conditions, ψW comes out to be zero, which is the highest potential of the water, and under unsaturated conditions, it is always expressed as –ve value. Under natural soil environment, soil moisture movement is mainly controlled by the hydraulic potential (ψh), which is the total moisture potential. There is a brief explanation regarding all the components of the soil moisture potential one by one.

2.1 Hydraulic potential

ψh is the total moisture potential, that is, ψt, which is the sum of other potentials by virtue of its pressure (ψp), attractive forces (ψm), and gravity (ψg) [14]. The ψh/ψt provides direction of the movement of soil moisture; however, if ψh is the same throughout the soil profile (under pounded conditions or under prolonged rainfall), then the water will not move at all in the soils as energy state is the same throughout and moisture only moves under the deviation in the moisture levels/ energy levels. Normally under the unsaturated soils, the water moves from the lesser to higher negative potential. Moisture potential of soil delineation is quite important, as it directs us irrigation timings [9, 14]. Further, hydraulic conductivity of a particular soil having a particular textural class is very important, which is further important for nutrient movements within the plants. The slope of the curve between flux (discharge area^{-1} time^{-1}) and hydraulic gradient decides the hydraulic conductivity itself varied with texturally divergent soils (**Figure 4**). This figure explains why movement of water differs in texturally divergent soils and we could manage our cultivation and management practices so as to increase the water-use efficiency.

2.2 Matric potential (ψm)

Different adsorption forces prevailing in the soil matrix are responsible for the ψm—the force of attraction of free water with soil particles [14]. The greater the adsorption forces, the more is the matric potential, and thus the water is less free. In other words, water is tightly attached to the soil particles. However, ψm is dependent on many factors, out of which soil texture is important, for example, sandy coarse-textured soils drained out moisture quickly at a smaller suction than clayey fine-textured soils because clayey soils have greater matric adsorption forces which

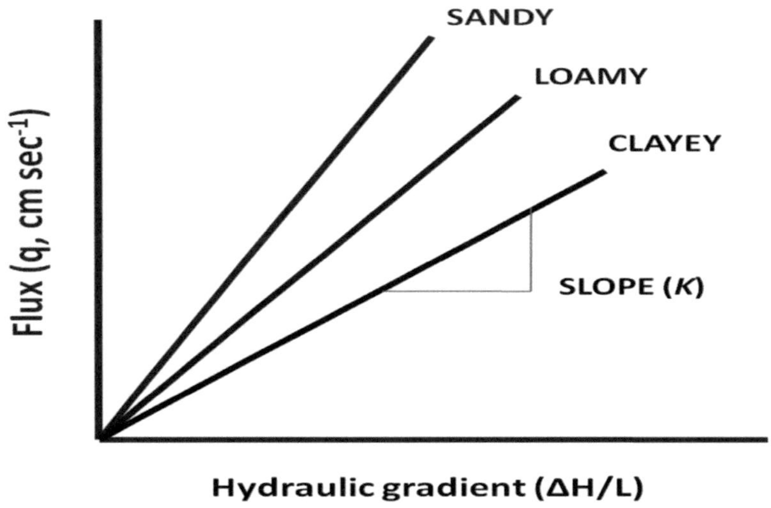

Figure 4.
Relationship between flux and hydraulic gradient in three texturally divergent soils.

hold the water tightly and not allowed the water to drain out quickly. In other words, clayey soil has more –ve values of ψm than that of sandy soils, depicting the higher capacity of former soil water holding capacity of clayey soils. Similarly, the soils with higher organic matter (OM) content have higher water content and thus greater –ve value. *It is very important to understand that the greater is –ve* ψm *value, the higher is the water content as water always moves from the higher potential to lower potential or from lesser –ve to more –ve as more negative values of* ψm *depict the lower water content.* ψm has been considered as capillary potential. If we consider weight as the unit for expressing the unit quantity of water, then ψm with respect to a particular height in the soil is the distance in the vertical direction between that selected height and level of water in a manometer. Generally, the ψm resulted from the two processes, namely, capillary "wedges" and "films," which cannot be changed without upsetting the others. K under saturated conditions varied in texturally divergent soils, due to attractive forces in soil separates and soil moisture (**Figure 5**). As shown in the picture, saturated hydraulic conductivity of sandy soil is more than of the clayey soil; however, the unsaturated conductivity of sandy soil decreases more steeply with increased suction and decreased from the clayey soils.

Ψm reported to be zero under saturated conditions; hence a –ve sign is always there under the unsaturated conditions which is the most prevalent situation in natural field conditions. Matric potential is always zero at the water level, positive below the water table, and negative above the water table. For measuring the suction or ψm in soils, we used tensiometer in soils (**Figure 6**) and set a particular reading for irrigating the fields.

However, tensiometer could measure the suction <0.85 bar (most prevalent in natural conditions), and pressure plate apparatus and tension plate assembly are used for measuring suctions >0.85 [7]. The graphical behavior of tension of soil moisture with absolute water content is developed through a soil moisture characteristic curve, which delineates the moisture levels that the soils could hold and thus helps in scheduling the irrigation to crops accordingly.

Under this scenario, the available soil moisture of Indo-Gangetic Plains is described by ψm [15]. Locally fabricated, low-cost tensiometers [16] that could delineate soil matric potential are generally preferred by the farmers for scheduling irrigation more particularly to rice [17, 18]. According to Kukal et al. [19],

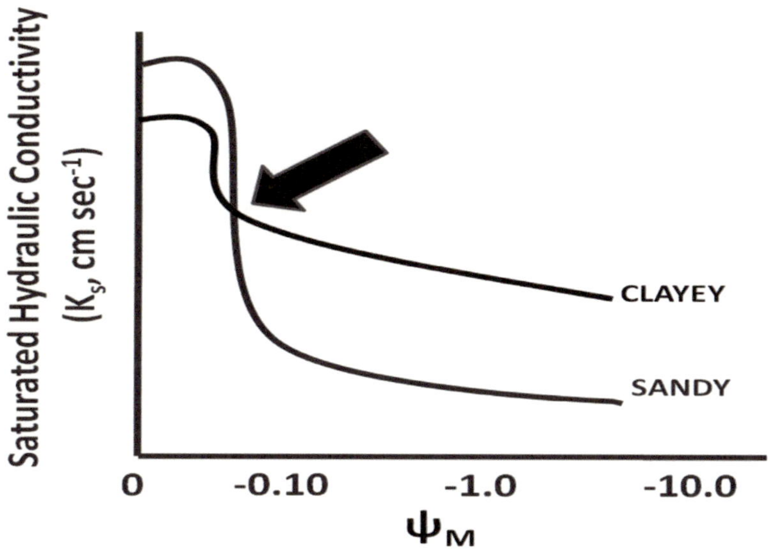

Figure 5.
Association between saturated hydraulic conductivity and ψ_m in two soils.

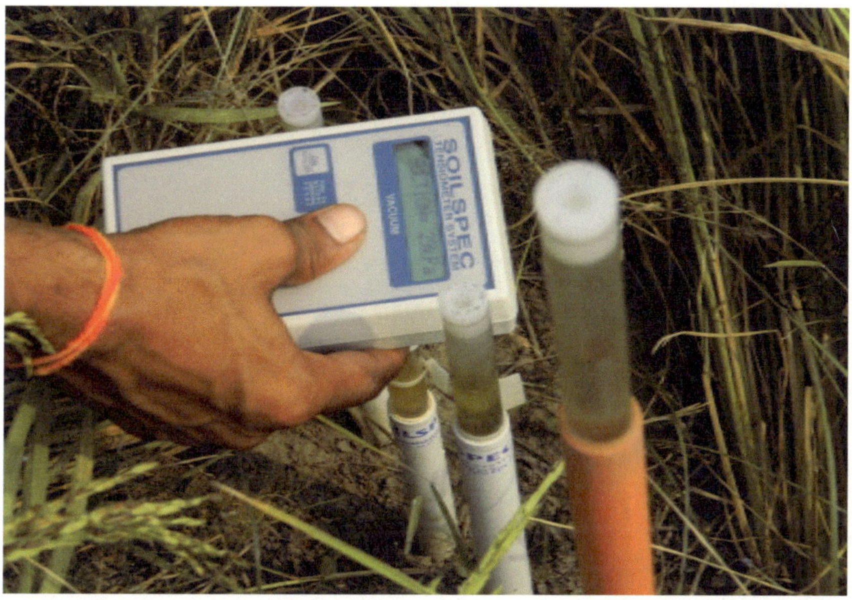

Figure 6.
Soil spec in action measuring soil matric potential [20].

increasing suction values to 2000 and 2400 $\pm$ 200 mm reduced the land productivity of the rice than earlier recommendation (2-day interval), which mean drying of soils to certain extent saves significant irrigation water without significantly affecting grain yields. Further, an average of 5-year study delineated (**Table 1**) a saving of up to 30% of irrigation water without adversely affecting the land productivity [21].

For measuring ψm, tensiometers are installed at 15–20 cm depth, because significant rhizosphere's portion of the rice crops retained to upper 15 cm [15], and therefore, tensiometers are placed at this depth, so that farmers could get the exact idea regarding the exact time to irrigate.

Year	% water saving	Yield differences
2006	29.6–30.7	+0.5–1.5%
2007	25–27.2	At par
2008	18–27.8	At par
2009	16.6–20.8	+0.5–1.0%
2010	11.1–21.4	At par

Source: Ref. [12].

Table 1.
Soil matric potential based irrigation water saving viz.-a-viz. yield differences.

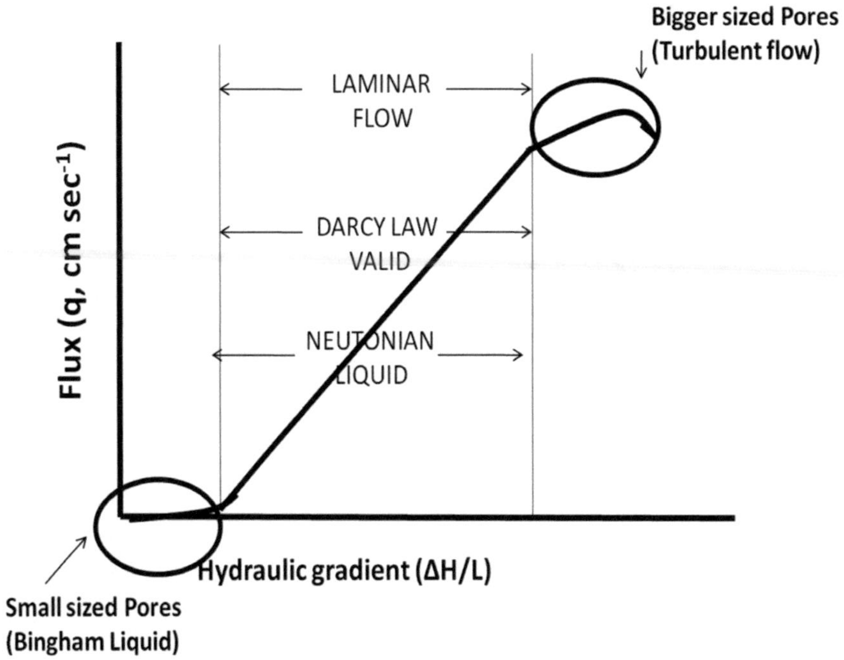

Figure 7.
Deviation in Darcy law.

2.3 Pressure potential

ψp is a vital constituent of soil moisture potential but under the saturated conditions which seldom exist in nature [22]. Generally, saturated conditions come only when rains up to a considerable duration or continuous irrigation. When saturated flow becomes high enough to be turbulent and lesser enough for not to generate any flux for a prolonged time is there to meet the constant drainage and evaporation and flow in these conditions is basically governed by the force of gravity but these conditions seldom exist in a field or under natural conditions as here all the soil pores are water filled and conducting it [8, 22]. Under this condition, the discharge is governed by Darcy's law, which further has some limitations as shown in **Figure 7**.

Negative pressure potential of unsaturated soil becomes positive in the saturated conditions and is delineated as submergence or pressure potential which is generally measured with a piezometer. A piezometer is a hallow tube open from both ends, passing from the reference point. If we consider weight as the unit of expression of the quantity of water, then certainly ψp is delineated by vertical space from the considered point and piezometer level of water, connected to that point in question. Pressure potential is always positive and zero below the water level and at and above the water level, respectively. *ψp and ψm are mutually exclusive to each other as if ψp is positive, then ψm is zero, while if ψm is negative, then ψp is zero.*

2.4 Gravitational potential

ψg constitutes an important soil moisture potential component which is not affected by the soil properties [14]. On considering weight as the unit of quantity of water, ψg comes out to be the vertical distance of elevation from a point under consideration to the point in question and is thus considered as the elevation distance from a point under consideration to the level of reference [14]. To raise an object against the gravitational force of attraction, some work must be done which is stored in the form of energy with respect to its gravity. Gravitational potential is zero at, positive above, and negative below the reference level. It does not depend upon soil properties; this is the reason why ψg is not considered while calculating the water potential. However, ψg played an important role and is considered while calculating the total water potential as

$$\psi t = \psi w + \psi g \tag{4}$$

Further, ψg is independent on the conditions of soil, water, weather, chemical, and pressure, while elevation levels are affecting it. Hence, height is the only criteria affecting the gravitational water in one and all [14].

2.5 Osmotic potential (ψs)

ψs is an important potential which is there in soil because of the salts in soil water and also due to the presence of the semipermeable layer, which only allowed water entry but not of the salts through it [14]. In soil-water interface, there are mainly two important semipermeable membranes, namely, air-water interface and cell wall in the roots. Air-water interface behaves near to the perfect semipermeable membrane, while cell wall of roots is not a perfect semipermeable membrane as it allows passage of salts as well as water through it. However, while studying liquid water flow in soils, ψs is an unimportant potential due to lack of semipermeable membrane in it, while in plants it is of much importance as plant ease to absorb water is greatly affected by ψs as the more the value of ψs, the higher the energy exerted by plant to pull deep underground water. Consider sodic/saline soil, through which the plants have to exert the water, and then it can exert a ψs equal to the permanent wilting point of soils. Thus determining the value of ψs = -RTCs, where R, T, and C_s represent universal gas constant (82 bars cm^{-2}), absolute temperature, and solute/salt concentration in soils, respectively, is the most difficult as it also includes those species which dissociate into the ions [9].

There are many terminological terms, namely, water-use efficiency at global and local levels and allocation efficiency pertaining to water used in the literature [20, 23] for sustainable use of the irrigation water throughout the globe. Further, Allan coined the term "virtual water" for human consumption. Further, published literature also delineate some terms pertaining to crop water, namely, green, blue, gray, and black water [20]. The most important term that pertains to human water use is referred to as "blue water" as it is rain water, which directly enters the lakes and is used by humans. For plants, the most important water term is "green water" as it is there in soil pores and meets the transpiration demands of plants to produce biomass [24]. Domestic activities such as bathing and dishwashing constitute the "gray water," while "black water" is the produce of laundry which consists of toilet water. Among all the different categories of water, only gray water has the huge potential of being reused, which further cut off the freshwater demand by 30% in cities [9].

3. Soil moisture balance computation

Computation of the soil water balance is an important aspect which needs to be focused, and their detailed methodological understanding is a must more particularly for the budding scientists. Nowadays, many research papers are published in the journals of repute, publishing effect of RCTs, namely, laser leveler, DSR, zero tillage, etc., on improving the water as well as land productivity without discussing much on the estimation part. Thus, there is confusion in between the scientists especially budding ones as to how to estimate the performance of a particular RCT under different conditions of soil texture and climate. Moreover, there is an interest in the evaluation of these RCTs in improving the production potentials by diverting maximum ET water to the T components, thereby providing higher nutrients to the plants [15, 25, 26] and recommending them as per the soil textural class as these technologies are location specific and not a single technology is capable of performing equally under all the conditions. Hence, there is a need to delineate the estimation/calculative part of the different moisture balance components of the soil.

Nowadays, agricultural scientists are focusing on techniques to reduce the soil evaporation [27–30] for partitioning higher part of the soil moisture from evaporation (unproductive component) to the transpiration (productive component) for improving the grain yields of the farmers of the water-stressed regions throughout the globe. Countries, namely, Switzerland, the USA, Germany, the Netherlands, Sweden, etc., recognized the significance of the aquifer management [29, 31]. Proper water allotment, as per demand and availability, is a decisive issue [29, 32]. Further, to feed 9.5 billion population up to 2050 [33], around 60% more food [34] is required to produce from the shrinking natural resources, namely, land and water [29, 35–37]. One other claimed way is to use waste or industrial water, but it needs efforts to clean it first which sometimes is not an easy step. Climate change further complicated the conditions as it has a significant effect on the agriculture by altering the rainfall patterns, CO_2 concentration, air temperature, etc. [29, 36, 38]. Improved standards of living [39] and altered eating habits [8], which need more consumption of water, make the scenario more complex. Therefore, a challenge in front of the agricultural scientist to come out from this situation seems to be a bit difficult. The only way is to partition greater fraction of evapotranspiration (ET) component share to the transpiration side for improving the land productivity even in the water-stressed region, but without knowing the proper procedure for calculating the evaporation component, the budding scientists will not able to assess the impact of different RCTs for this partition. Therefore, estimation of the different soil moistures/water balance components is a must and of course very important for having an idea to what are the added water amounts (through rainfall or irrigation) and what are the lost amounts (either through evaporation, transpiration, seepage, drainage, change in profile moisture storage, etc.). Among all the water lost components on the left side, evapotranspiration generally denoted by ET is most important whose share remained almost the same [29, 38]. Further among ET, E pertains to unproductive water from open surfaces which must be partitioned to T for having higher yields [8, 20]. However, water loss through D and S is always away from the rhizosphere and thus is not used by the crop plants for meeting their ET requirements.

Before sowing and after harvesting the crop, namely, during the intervening periods, profile moisture storage change could be measured, which further played an important role in the cultivation of fodder crops. A soil water balance component provides a way out to identify technologies which improve water productivity. Up to now, this period is the least attended as results of applied treatments evaluated are analyzed during this period [20, 27, 40, 41]. However, the intervening period delineation of soil moisture dynamics helped to assess the residual effects of these RCTs applied during the main crop [40, 41]. Therefore, for sustainable and judicious use of irrigation water, the analysis of the soil water balance component is very important. The following are the important parameters of the soil water balance which needs to be calculated for evaluating the performance of any RCT in any region of the globe:

$$E + T + D + S + \Delta G = R + I \tag{5}$$

where E is the evaporation, T is the transpiration, D is the drainage, S is the seepage, ΔG is the profile moisture change, R is the rainfall, and I is the irrigation. Details along with their calculative/instrumental part are discussed below.

3.1 Rainfall (R)

Rainfall is an important soil water balance component which decides the fate of the rainfed crops grown particularly in the submountainous tracts where there is no irrigation facilities, which might be because of the hard subsurface and very deep underground water table [17, 23]. Therefore, its timely quantification is very important for recognizing stressed areas which further helps in rescheduling irrigation plans for improving land and water productivity over here. Received rainfall is estimated using a rain gauge, which is installed permanently at the location/period of experimentation, which is further used in calculating the rainfall water productivity (WP_I) [15]. However, one should be very careful that the spot selected for rain gauge installation should be away from huge buildings or any obstacles or any hindrance. Necessary correction factor must be applied, which is the case of the heavy rainfall if rain gauge's cylinder overflowed [39]. Many times, it is observed that rain gauge base is not fixed, which may result in tilting of the gauge while recording the rainfall; thus while installing it, it should be made sure that it should be fixed by using cement and sand mixture, so that no error in calculations will be there [15, 23, 39].

3.2 Irrigation water amount (I)

Irrigation is the most important for having potential agricultural yields in any area. But generally irrigation water-use efficiency is quite low in spite of the fact that water already is a limiting factor. Further, irrigation is an important input component for soil water solution; however, its exact measurement is generally not there, even in water management experiments. Nowadays we are well equipped with the water measuring meters which accurately measured the water amount which is being applied to a particular plot under any treatment, namely, area velocity flow meter (AVFM 5) which provides a digital reading of water supplied in any plot [15]. Generally, irrigation water depth of 50 and 75 mm in wheat and rice plots supplied which could be measured through the sensor (fitted in the pipe through which water enters a particular plot) of AVFM [27]. GREYLINE is the company manufacturing the Digital flow meter (**Figure 8**) the irrigation water measuring irrigation water device on a quantitative basis. Their sensor has to be fit

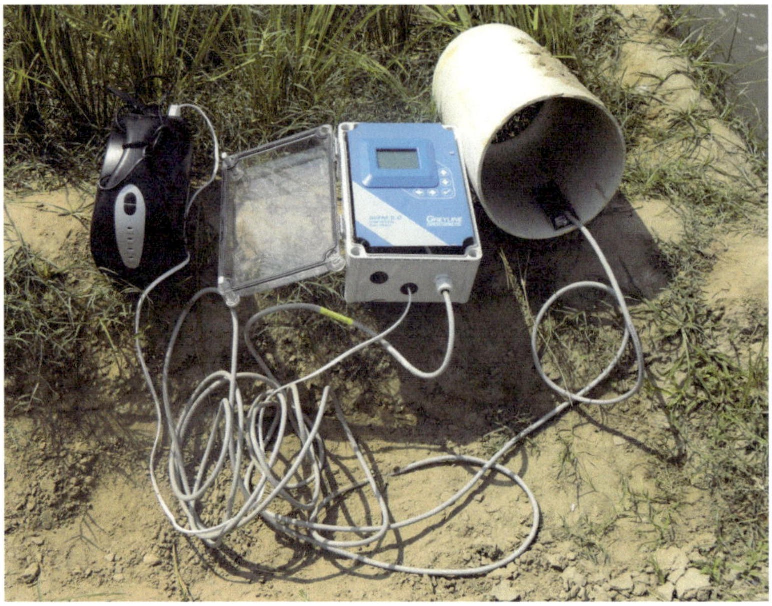

Figure 8.
Area Velocity Flow Meter for calculating the irrigation water applied.

in the plastic pipe. When water applied to a particular plot equipped by a particular treatment, sensor placed in the pipe starts recording and displaying the quantity of water entered in the plot in liters which could be further be used in calculating the irrigation water productivity of differently treated plots. As there is no electric supply in the remote agricultural fields, hence a battery is required for its power. Further, calibrations are required before using it by filling the water in a Known volume of drum and in case of any discrepancy, a correction factor must be applied for further calculations for applied irrigation amounts in the agricultural water management experiments so that correct irrigation water productivities will be delineated under different treatments.

3.3 Evaporation (E)

Evaporation generally is known as the unproductive loss of the water from any surface, namely, soil or water or leaf, when liquid water changed to vapor form in the presence of the certain energy, and is affected by establishment methods [8, 12, 20, 29, 42] as mulched plots experienced lesser evaporation losses. However, through the stomata of the leaf, loss of liquid water to atmosphere in gaseous form coined as productive loss, delineated as "transpiration (T)" as under transpiration pull along with water nutrient also enters into the plants through the roots which further results in higher grain yields. Therefore, for having higher production of the plants, higher transpiration is required; thus, every effort is made to divert a greater fraction of the ET share of the soil moisture to the T component [15, 39]. Generally, lysimeters are used for delineating the evaporation, while transpiration is delineated after subtracting other water loss factors from rainfall + irrigation. Lysimeters [41, 42] comprised of two pipes of PVC, the outer (0.16 m) being wider than the inner (0.102 m) in diameter while both of the same length (0.20 m). Porous end cap is used to seal the inner one from downward side, while the outer one was opened from both sides for making soil environment homogeneous in mini-lysimeter and the outer field. Cylindrical auger is used for making space in the field for fitting wider outer pipe (0.20 m long), in which inner soil-filled pipe (duly closed from downside with an end caps) is placed.

The inner PVC tube weight was measured daily at 0.900 hours (**Figure 9d**) using a digital weighing balance. Mini-lysimeters are used (**Figure 9a–d**) in the treatment plots, where daily evaporation needs to be worked in mm below the crop canopy [20, 42, 43]. Providing permanent location in the field plots receiving differential treatments throughout the season is the main objective of providing outer PVC pipe, where evaporation could be regularly measured. Hammer is used for inserting the narrower inner PVC pipe in the field during each sampling (**Figure 9a**), which removed from the plot with the help of chain-pulley arrangement (**Figure 9b**). Weeds growing on the mini-lysimeters must be cut and removed, so that it may not affect evaporation readings. Without any soil disturbance, inner pipes should be placed in the outer pipes, and daily in the morning, about 9:00 am, lysimeters were weighed (**Figure 9d**) and placed back in the outer PVC pipe.

3.3.1 Delineation of calculations of evaporation

Mostly, very little discussion is there in different research papers regarding the calculative part of the evaporation. Hence, the repetition of the carried-out work under differentially textured soils/agroclimatic conditions is quite difficult. As far as the calculative part, different lysimeters were installed in different plots receiving differently established methods/techniques.

Figure 9.
Step-wise technique of evaporation delineation with mini-lysimeters (a) Fitting of lysimeter in experimental plot, (b) use of chain-pulley for removing it from plots, (c) removed lysimeter, (d) weighing of lysimeters within the plots receiving differential treatments [23].

Let us suppose.
Day 1 (Mass of the Lysimeter + Soil) = A g.
Day 2 (Mass of the Lysimeter + Soil) = B g.
Evaporated moisture mass after 1 day = A-B = X g (suppose it is 15 g).
1 g = 1 cm^3 (15 g = 15 cm^3).

For calculating evaporated water in 24 hours under differential treatments, the differential lysimeter weight in cm^3 needs to be divided by the lysimeter area (πr^2) cm^2 where r is the radius. Let us suppose radius was 7.5 cm.

Hence, evaporated water = 15 cm^3/3.14 × 7.5 cm × 7.5 cm = 0.085 cm.

Delineation of moisture evaporated from a particular treatmental plot during the last 24 hours is quite important. The "cm" units are converted into "mm" by multiplying it by 10. Therefore, in the above case, 0.85 mm (0.085 × 10 = 0.85 mm) of evaporation is there. With this way, the performance of different RCTs in reducing evaporation and thereby promoting the transpiration could be delineated in a particular region (**Figure 10**).

3.4 Drainage (D)

Drainage is the loss of irrigation or rain water in the downward direction beyond the rhizosphere. Therefore, drained away water could never be used up by the plants. Hence it needs to be checked for providing more moisture to the rhizosphere. In wheat, generally, drainage losses are assumed to be negligible or near to 100 mm, while in the rice season, drainage losses are of significance (>2000 mm). For calculating the drainage losses in the rice season, electronic tensiometers are installed at 450 and 600 mm assuming rhizosphere up to 500 mm [15].

For a drainage calculation, unsaturated hydraulic conductivity needs to be delineated by using the disk permeameter, which is used throughout the soil profile (**Figure 11**).

Now, for calculating the flux using Darcy's law (Eq. (6)), delineation of the unsaturated K of the transitional layer on a daily basis is very important, which is further expressed as deep drainage.

$$q = K.\Delta H/L \quad (6)$$

where Q is the flux, K is the unsaturated hydraulic conductivity, and ΔH/L is the hydraulic gradient.

Hydraulic gradient (ΔH/L) changed to the suction gradient (ΔΨt/L), for tensiometers

$$q = K.\Delta\Psi t/L \quad (7)$$

where Ψt is the total potential which is the sum of matric and gravitational potentials, namely, Ψm + Ψg, which are delineated as in cm and kPa, respectively. kPa is easily converted into cm by multiplying it with 10. Disk permeameter (**Figure 11**) is generally used for estimating unsaturated hydraulic conductivity values up to 0–150 cm. For estimating water drained deep through the soil profile, Eq. (4) is used.

$$q = K.\Delta\Psi t/L \quad (8)$$

$$q = K.\Delta\Psi A - \Delta\Psi B/L \quad (9)$$

$$q = K.\{(\text{Tensiometer readings at 45cm} - 45) - (\text{Tensiometer readings at 60cm} - 60)/15 \quad (10)$$

Figure 10.
Working of electronic tensiometers (a) Fitting of tensiometers in the field, (b) filling of water in tensiometers, (c–e) installed tensiometers, (f) measuring of matric potential using digital soil spec [24].

Sometimes under field conditions, different length tensiometers had to be used depending upon their availability; hence, a correction factor is applied to nullify this effect.

$$\text{Correction Factor} = \text{Tensiometer reading} – 9.8 * (\textit{Tensiometer length}\ \text{row number}/100) \tag{11}$$

Generally, the tensiometer reading is in kPa, but for the soil water balance studies, readings in "cm" are necessary, which are converted by multiplying kPa reading with 10. After filling reading in Eq. (5), flux (q)/drainage loss in different plots could be easily delineated.

Figure 11.
Disk permeameter for delineation of un-saturated hydraulic conductivity [23].

3.5 Seepage (S)

S is the sideway water travels from side to side of the bunds, which could alter the water amounts used. For delineating the seepage loss in the rice season, water level variation in whole plots and infiltration rings is recorded during every irrigation [15, 43]. After each irrigation/heavy rainfall, seepage was calculated. After 2–3 hours depending upon the soil textural class, water from plot disappears, and then, the ring water level provides us with a scheme of the seepage losses from a particular experimental plot.

3.6 Change in profile moisture (ΔG)

Profile moisture change is also an important part of the soil water balance equation. For measuring soil profile moisture change, the thermogravimetrical method is used for measuring moisture before sowing and after harvesting throughout the profile up to a depth of 1.5 m.

$$\text{Moisture of soil } (\text{g g}^{-1}) = \text{Fresh soil mass (g)–oven-dried soil mass (g)/ Oven-dried soil mass (g).} \tag{12}$$

From the above conversion, above weight basis (g g^{-1}) values of soil moisture to volumetric basis ($\text{cm}^3 \text{ cm}^{-3}$), these values must be multiplied with a respective bulk density.

$$Øi = W \times D_b \tag{13}$$

where Øi is the volumetric soil moisture ($cm^3\ cm^{-3}$); W is the mass basis soil moisture; and D_b is the bulk density.

For D_b determination, generally, core method [22] was used. Under this method, undisturbed metallic soil cores are used for calculating the D_b, and fresh core weight was measured. Then, fresh soil + cores weight was recorded, and then, both fresh soil and cores are dried for 1 day in an oven at 105°C. For D_b, the dried weight of soil is divided with the internal volume of the metallic cores [15]. Further for a specific depth under consideration, moisture (cm) is determined by

$$\text{In a specific depth of soil, soil moisture (cm)} = \text{Øi} \times \text{soil profile depth} \quad (14)$$

Further, for delineating soil profile moisture up to 150 cm, each depth value of soil moisture is added up to have soil profile moisture (cm), which is further multiplied by 10 to get soil moisture of the whole profile in mm, the required units for the soil moisture balance.

By adopting above methodology for calculating different soil moisture components, namely, rainfall, irrigation, evaporation, transpiration, seepage, drainage, and change in profile soil moisture, one could easily delineate the soil moisture components or validate the performance of a particular resource conservation technology, namely, happy seeder, laser leveler, tensiometers, direct-seeded rice, etc., in improving the yield potentials by partitioning the maximum share of the evapotranspiration water from evaporation to transpiration.

4. Conclusion

Underground water is globally declining down which in itself is a matter of great concern. Further, population pressure is rising day by day whose requirements whether of food, fiber, etc. should be met out from the ever-diminishing resources, namely, water and land. Climate change further complicated the whole scenario by one or other way. Thus, under this whole current scenario, it is very much important to first have knowledge regarding soil moisture movement under the impacts of different soil moisture potentials, namely, matric potential, solute potential, and gravitational potential, so that irrigation water is applied as required for having higher water-use efficiency for which tensiometers may serve the purpose under the field conditions. Further, many RCTs are being proposed in the water-stressed regions for establishing the wheat-rice cropping sequence with claim to have higher water-use efficiency and, thus, higher land and water productivity. But a careful observation delineates that all of these RCTs are not universally applicable; rather their performance varied as per differential sand, silt, and clay ratios, soil slope, and agroclimatic conditions. Therefore, the first idea regarding different soil water potentials and then, secondly, rechecking of different recommended RCTs in a diversion of maximum share of green water from E to T are required. For this, estimating different soil moisture balance components and therefore their instrumental/calculative part needs more attention in the budding scientists more particularly dealing with the agricultural water management experiments in the water-stressed regions of the globe.

Conflict of interest

No conflict of interest is expressed by the authors.

Abbreviations

D_b	bulk density
RCT	resource conservation technologies
Es	evaporation from soil surface
D	drainage
Ψm	potential by virtue of attraction due to soil matrix
Ψg	potential by virtue of gravity
W	mass basis moisture
ΔH/L	hydraulic gradient
Øi	moisture content on volumetric basis
SMP	soil matric potential
q	flux
K	unsaturated hydraulic conductivity
I	irrigation
T	transpiration
AVFM	area velocity flow meter
PVC	polyvinyl chloride
DSR	dry-seeded rice
ET	evapotranspiration
R	rainfall
WP_I	irrigation water productivity
E	evaporation
S	seepage

Author details

Rajan Bhatt[1]* and Ram Swaroop Meena[2]

1 Regional Research Station, Kapurthala, Punjab Agricultural University, Ludhiana, Punjab, India

2 Institute of Agricultural Sciences, Banaras Hindu University, Varanasi, India

*Address all correspondence to: rajansoils@pau.edu

References

[1] Dane JH, Topp CG, Young MH, Sisson JB. Tensiometry. SSSA Book Series. 2002;**5**(4):575-608. DOI: 10.2136/sssabookser5.4.c21

[2] Gallopin GC, Rijsberman F. Three global water scenarios. International Journal of Water. 2000;**1**(1):16-40

[3] Yadav S, Evangelista G, Faronilo J, Humphreys E, Henry A, Fernandez L. Establishment method effects on crop performance and water productivity of irrigated rice in the tropics. Field Crops Research. 2014;**166**:112-127

[4] Yadav AS, Upadhyay RK. Climate smart agriculture. In: Rawat SK, Narain S, editors. Agricultural Technology for Sustaining Rural Growth. New Delhi, India: Biotech Books; 2020. pp. 69-83

[5] Cosgrove WJ, Loucks DP. Water management: Current and future challenges and research directions. Water Resources Research. 2015;**51**: 4823-4839. DOI: 10.1002/2014WR016869

[6] Alberto MCR, Wassmann R, Hirano T, Miyata A, Hatano R, Kumar A, et al. Comparisons of energy balance and evapotranspiration between flooded and aerobic rice fields in the Philippines. Agricultural Water Management. 2011;**98**:1417-1430

[7] Hira GS, Singh R, Kukal SS. Soil matric suction: A criterion for scheduling irrigation to rice (Oryza sativa). Indian Journal of Agricultural Sciences. 2002;**72**:236-237

[8] Mahajan A, Gupta RD. The Rice–wheat cropping system. In: Integrated Nutrient Management (INM) in a Sustainable Rice—Wheat Cropping System. Dordrecht: Springer; 2009. pp. 109-117. DOI: 10.1007/978-1-4020-9875-8_7

[9] Buckingham E. Studies on the movement of soil moisture. Bureau of Soils. Washington, DC, USA: Department of Agriculture; 1907

[10] Gardner W. The capillary potential and its relation to soil moisture constants. Soil Science. 1920;**10**: 357-359

[11] Richards LA. Capillary conduction of liquids through porous mediums [PhD thesis]. Cornell University; 1931

[12] Mathew RP, Feng Y, Githinji L, Ankumah R, Balkcom KS. Impact of no-tillage and conventional tillage systems on soil microbial communities. Applied and Environmental Soil Science. 2012; **2012**:1-10. DOI: 10.1155/2012/548620

[13] Gardner-Outlaw E, Gardner-Outlaw T, Engelman R. Sustaining Water, Easing Scarcity: A Second Update. Washington, DC: Population Action International; 1997. p. 19

[14] Gupta N, Yadav S, Humphreys E, Kukal SS, Singh B, Eberbach PL. Effects of tillage and mulch on the growth, yield and irrigation water productivity of a dry seeded rice-wheat cropping system in North-West India. Field Crops Research. 2016;**196**:219-236

[15] Alberto MCR, Wassmann R, Hirano T, Miyata A, Hatano R, Kumar A, et al. Comparisons of energy balance and evapotranspiration between flooded and aerobic rice fields in the Philippines. Agricultural Water Management. 2011;**98**:1417-1430

[16] Studer C, Spoehel S. Potential and actual water savings through improved irrigation scheduling in small-scale vegetable production. Agronomy. 2019; **9**(12):888. DOI: 10.3390/agronomy9120888

[17] Shock CC, Feibert EBG, Seddigh M, Saunders LD. Water requirements and

growth of irrigated hybrid poplar in semi-arid environment in eastern Oregon. Western Journal of Applied Forestry. 2002;**17**:46-53

[18] Shock CC, Feibert EBG, Saunders LD. Irrigation criterion for drip irrigated onions. Horticultural Science. 2003;**35**:63-66

[19] Kukal SS, Hira GS, Sidhu AS. Soil matric potential-based irrigation scheduling to rice (Oryza sativa). Irrigation Science. 2005;**23**:153-159

[20] Bhatt R, Kukal SS. Soil moisture dynamics during intervening period in rice-wheat sequence as affected by different tillage methods at Ludhiana, Punjab, India. Soil and the Environment. 2015a;**34**(1):82-88

[21] Karimi H, Ghassemzadeh HR, Pashaee P, Shahamat EZ. Proposing an appropriate soil water content estimation technique for Iran. Agricultural Engineering International: CIGR Journal. 2015;**17**(2):1-10

[22] Bodman GB. Nomograms for rapid calculation of soil density, water content and total porosity relationship. American Society of Agronomy. 1942; **34**:883-893

[23] Bhatt R, Kukal SS. Moisture retention trends during the intervening period of differently established rice-wheat cropping pattern in a sandy loam soil. International Journal of Farm Sciences. 2014;**4**(2):7-14

[24] Bhatt R, Kukal SS. Delineating soil moisture dynamics as affected by tillage in wheat, rice and establishment methods during intervening period. Journal of Applied and Natural Science. 2015b;**7**(1):364-368

[25] Falkenmark M. Coping with water scarcity under rapid population growth. In: Conference of SADC Minis-Ters. Pretoria: Elsevier; 1995. pp. 23-24

[26] Humphreys E, Kukal SS, Christen EW, Hira GS, Singh B, Yadav S, et al. Halting the groundwater decline in north-West India-which crop technologies will be winners? Advances in Agronomy. 2010;**109**: 156-199. DOI: 10.1016/S0065-2113(10)09005-X

[27] Konikow LF. Long-term groundwater depletion in the United States. Groundwater. 2015;**53**:2-9

[28] Lund J, Harter T. California's groundwater problems and prospects. Available from: https://californiawaterblog.com/2013/01/30/californias-groundwater-problems-and-prospects/ [Accessed: 11 March 2016]

[29] Scanlon BR, Keese KE, Flint AL, Flint LE, Gaye CB, Edmunds WM, et al. Global synthesis of groundwater recharge in semiarid and arid regions. Hydrological Processes. 2006;**20**: 3335-3370

[30] Werner AD, Ward JD, Morgan LK, Simmons CT, Robinson NI, Teubner MD. Vulnerability indicators of sea water intrusion. Ground Water. 2012;**50**(1):48-58. DOI: 10.1111/j.1745-6584.2011.00817.x

[31] Xu Y, Beekman HE. Review of groundwater recharge estimation in arid and semi-arid Southern Africa. 2003. Available from: http://unesdoc.unesco.org/images/0013/001324/132404e.pdf [Accessed: 11 March 2016]

[32] Hamaiedeh ALH, Bino M. Effect of treated grey water reuse in irrigation on soil and plants. Desalination. 2010;**256**: 115-119. DOI: 10.1016/j.desal.2010.02.004

[33] Hoekstra AY, Hung PQ. Globalization of water resources: International virtual water flows in relation to crop trade. Global Environmental Change. 2005;**15**(1): 45-56. DOI: 10.1016/j.gloenvcha.2004.06.004

[34] FAO. FAOSTAT database: Agriculture production. In: Food and Agriculture Organization of the United Nations. Rome: Food and Agriculture Organization of the United Nations; 2013

[35] Scanlon BR, Jolly I, Sophocleous M, Zhang L. Global impacts of conversions from natural to agricultural ecosystems on water resources: Quantity versus quality. Water Resources Research. 2007;**43**:210-219. DOI: 10.1029/2006WR005486 W03437

[36] Singh R, Singh GS. Traditional agriculture: A climate-smart approach for sustainable food production. Energy, Ecology and Environment. 2017;**2**(5): 296-316. DOI: 10.1007/s40974-017-0074-7

[37] Nace RL. Water resources: A global problem with local roots. Environmental Science & Technology. 1967;**1**:550-560

[38] Singh A, Panda SN. Integrated salt and water balance modelling for the management of waterlogging and salinization. 2: Application of SAHYSMOD. Journal of Irrigation & Drainage Engineering. 2012c;**138**(11): 964-971

[39] Guan D, Zhang Y, Kaisi MMA, Wang Q, Zhang M, Li Z. Tillage practices effect on root distribution and water use efficiency of winter wheat under rain-fed condition in the North China plain. Soil and Tillage Research. 2015;**146**:286-295

[40] Panwar A, Renner M, Kleidon A. Imprints of evaporation and vegetation type in diurnal 2 temperature variations. Hydrology and Earth System Sciences. 16 March 2020. DOI: 10.5194/hess-2020-95

[41] Hoekstra JM, Boucher TM, Ricketts TH. Confronting a biome crisis: Global disparities of habitat loss and protection. Ecology Letters. 2015;**8**: 23-29. DOI: 10.1111/j.1461-0248.2004.00686.x

[42] Gupta N, Eberbach PL, Humphreys E, Singh B, Yadav S, Kukal SS. Estimating soil evaporation in dry seeded rice and wheat crops after wetting events. Agricultural Water Management. 2019;**217**:98-106. DOI: 10.1016/j.agwat.2019.02.037

[43] Singh B, Eberbach PL, Humphreys E, Kukal SS. The effect of rice straw mulch on evapotranspiration, transpiration and soil evaporation of irrigated wheat in Punjab, India. Agricultural Water Management. 2011; **98**:1847-1855

Chapter 2

Soil and Water Conservation Measures for Agricultural Sustainability

Anita Kumawat, Devideen Yadav, Kala Samadharmam and Ittyamkandath Rashmi

Abstract

Limited natural resources are available on the planet under immense pressure due to the ever-increasing population and changing climate. Soil and water are fundamental natural resources for the agricultural production system. Anthropogenic and adverse natural activities are the major factors for the deterioration of natural resources. Among the various degradation processes, soil erosion is one of the serious threats for the deterioration of soil and water resources. In India, about 68.4% of the total land area has been degraded by the water erosion process. Intensive agricultural practices accelerate the soil erosion process. Similarly, increased exploitation of groundwater resulted in depletion of groundwater level. Hence, the holistic management of soil and water resources is indispensable for agricultural sustainability as well as for the protection of the natural ecosystem. Development and adoption of improved technologies, judicious use of natural resources, and effective management practices are the need of the hour for protection of soil and water from degradation. This chapter highlights the status of natural resource degradation, erosion processes and, soil and water conservation strategies for agricultural sustainability and soil health in the long run.

Keywords: agricultural sustainability, conservation measures, erosion, natural resources, soil, water

1. Introduction

Soil and water are indispensable for the existence and survival of all terrestrial life. These are the basic resources to the requirement for food, feed, fuel, and fiber of human beings. Soil supports plant life by providing a medium for their growth and development [1, 2]. It is a non-renewable natural resource and susceptible to rapid degradation through various forms of erosion processes. Worldwide, around 52% of total productive land has been degraded by various kinds of degradation processes and almost 80% of the terrestrial land is affected by water erosion [3, 4]. Further, annually ~10 million hectares (mha) of cropland becomes an unproductive at the global level due to soil erosion with an average rate of 30 t ha^{-1} $year^{-1}$ soil erosion [5]. It has been estimated that water erosion results in a global flux of sediments of 28 Pg $year^{-1}$ [6]. This, extensive degradation of finite soil resources can severely jeopardize global food security while deteriorating environmental quality. On the other hand, the future of living beings and agricultural production systems is at stake due to continuously depleting aquifers and increasing pressure on underground water under projected climate change scenarios [7]. Moreover, climate change will increase water demand globally by about 40% of the water needed for irrigation [8]. Hence, under the emerging scenario of acute water shortages and land degradation, we must focus our effort on the development and adoption of efficient approaches for soil and water conservation as well as for agricultural sustainability. Even the theme for "World soil day," 2019 was "stop soil erosion, save our future" to raise awareness on the importance of sustaining healthy ecosystems and human well-being. Judicious use and management soil and water resources are more vital now than ever before to satisfy the needs of the ever-growing world population [9]. Conservation of soil and water has several agronomic, environmental, and economical benefits. Worldwide, around US$ 400 billion annual cost of on- and -off-site erosion has been estimated for replenishing lost nutrients, cleaning of water reservoirs and conveyances, and preventing erosion [10, 11].

2. The extent of land degradation

Globally, changes in land use and management practices accelerated soil erosion and have led to irrevocable land degradation, which is affecting 23.5% of the earth's land area [12, 13]. Soil erosion is one of the serious problems which not only impair the quality of land and water resources but also harm agricultural production and the socio-economic condition of farmers. Soil erosion has degraded about 32% of total land area in the USA, 30.7% in China, 16% in Africa, 17% in Europe, and 45% in India through a wide range of degradation processes [14]. Among various land degradation processes, water erosion is a major problem affecting 68.4% of the total land area in India [15, 16]. In India, various organizations have estimated the extent of land degradation (**Table 1**). NBSS and LUP has been reported about 146.8 mha degraded land area in India [17].

A harmonization exercise was done involving various organizations, to work out the water erosion, wind erosion, physical, and chemical degradation in India [18].

Agency	Estimation year	Degraded area (mha)
National Commission on Agriculture	1976	148
Ministry of Agriculture-Soil and Water Conservation Division	1978	175
Department of Environment	1980	95
National Wasteland Development Board	1985	123
Society for Promotion of Wastelands Development	1984	130
National Remote Sensing Agency	1985	53
Ministry of Agriculture	1985	174
Ministry of Agriculture	1994	107
National Bureau of Soil Survey and Land Use Planning (NBSS&LUP)	1994	188
NBSS&LUP (Revised)	2004	147

Table 1.
Extent of land degradation estimated by different agencies in India.

Degradation type	Arable land (mha)	Open forest* (mha)	Data source
Water erosion (>10 t/ha/year)	73.27	9.30	ICAR-IISWC
Wind erosion (Eolian)	12.40	—	ICAR-CAZRI
Sub-total	85.67	9.30	
Chemical degradation			
Exclusively salt-affected soils	5.44	—	ICAR-CSSRI, NBSS&LUP and NRSA, 2004
Salt-affected and water eroded soils	1.20	0.10	
Exclusively acidic soils#	5.09	—	NBSS&LUP, 2005
Acidic and water eroded soils#	5.72	7.13	
Sub-total	17.45	7.23	
Physical degradation			
Mining and industrial waste	0.19		Visual interpretation of satellite data, NRSA, 2003
Permanent Water logging$	0.88		
Subtotal	1.07		
Total	104.19	16.53	
Grand total (Arable + open land)	120.72		

**Area with <40% tree canopy cover.*
#pH < 5.5 and areas under paddy and plantation crops were also included in the total acid soils.
$Sub-surface water logging is not considered.

Table 2.
Harmonized data of degraded and wastelands in India.

The harmonized data on degraded and wastelands with all possible combination classes is given in **Table 2**.

3. Soil Erosion and erosion causing agents

Soil erosion is the removal of topsoil by the physical forces of erosion causing agents at a greater rate than the rate of its formation. Initially, erosion removes the nutrient-rich fertile top layer of soil which leads to the reduced production potential of soil. Soil erosion is classified into two categories, i.e., accelerated and geological erosion. Geological erosion is the natural phenomenon, occurs through the constant process of weathering and disintegration of rocks in which the rate of erosion remains lower than the soil formation rate. In contrast, in accelerated erosion, the rate of soil erosion exceeds a certain threshold level and becomes rapid. Anthropogenic activities such as slash-and-burn agriculture, overgrazing, deforestation, mining, and intensive and faulty agriculture practices are accountable for accelerated soil erosion [9]. This higher rate of soil erosion leads to the removal of organic matter and plant nutrients from the fertile topsoil and eventually lowering crop productivity. Hence, the conservation and management of natural resources are essential. Although the soil erosion cannot be eliminated, however it must be reduced to the level that can minimize its adverse impact on productivity and agricultural sustainability.

Water and wind are two key agents that degrade soils through various kinds of erosion processes. Globally, around 1100 mha is affected by water erosion (56% of the total degraded land) and around 28% of the total degraded land area is affected by wind erosion [19]. Runoff removes the soil particles from sloping and bare lands while the wind blows away loose and detached soil particles from unprotected lands. Other processes of land degradation are soil compaction, waterlogging, acidification, alkalinization, and salinization depends on parent material, climatic conditions, and crop management practices. In this chapter, we will discuss about the soil erosion by water, different types, processes, factors, and management.

4. Water erosion

Worldwide, water erosion is the most severe type of soil erosion. In this form of erosion, detachment, and transportation of soil particles from their parental source take place by water through the action of rainfall, runoff, hailstorm, and irrigation. Water erosion is a prevailing form of erosion in humid and sub-humid agro-ecosystems. It also creates the problem in arid and semiarid regions, characterized by an intensive rainstorm and scanty vegetation cover. Water erosion comprises three basic phases, i.e., detachment, transportation, and deposition. Rainfall is one of the major factors which causes the movement and detachment of soil particles. The detached soil particles seal the open-ended and water-conducting soil pores, reduce water infiltration, and cause runoff. The first two phases determine the quantity of soil to be eroded and the third phase determines the distribution of the eroded material along the landscape. If there is no dispersion and transport of soil particles, there will be no deposition. Hence, detachment and transport of soil particles are the primary processes of soil erosion. Understanding the mechanisms and extent of water erosion is crucial to manage and develop erosion control practices. Splash, sheet, rill and gully erosion are main forms of soil erosion by water (**Figure 1**). The other forms of water erosion are ravine formation, slip, tunnel, stream bank, and coastal erosion [20, 21]. The different forms of water erosion are described below:

4.1 Splash erosion

Splash erosion is the first form of soil erosion by water. Falling raindrops on the soil surface break the soil aggregates and disperse and splash soil particles from their source, known as splash erosion. The process of splash erosion involves raindrop impact on soil particles, a splash of soil particles, and the formation of craters [22]. The raindrops falling on soil surface act like a small bomb which disintegrates soil particles and forms cavities of contrasting shapes and sizes. The depth of craters is equal to the depth of raindrop penetration which is a function of raindrop velocity, size, and shape. In this form, soil particles can move only a few centimeters away from their source.

4.2 Sheet erosion

This is the next phase to splash erosion, which promptly initiates sheet erosion. The fertile topsoil surface is removed uniformly as a thin layer from the entire sloping surface area of the field by runoff water. Sheet erosion is a function of particle detachment, rainfall intensity, and land slope. The shallow flow of runoff water causes this type of soil erosion in which small rills are formed. This is the most common and severe form of soil erosion from an agricultural point of view as it removes the nutrient-rich top layer of soil. Out of total soil erosion, nearly 70% is caused by splash and sheet erosion only.

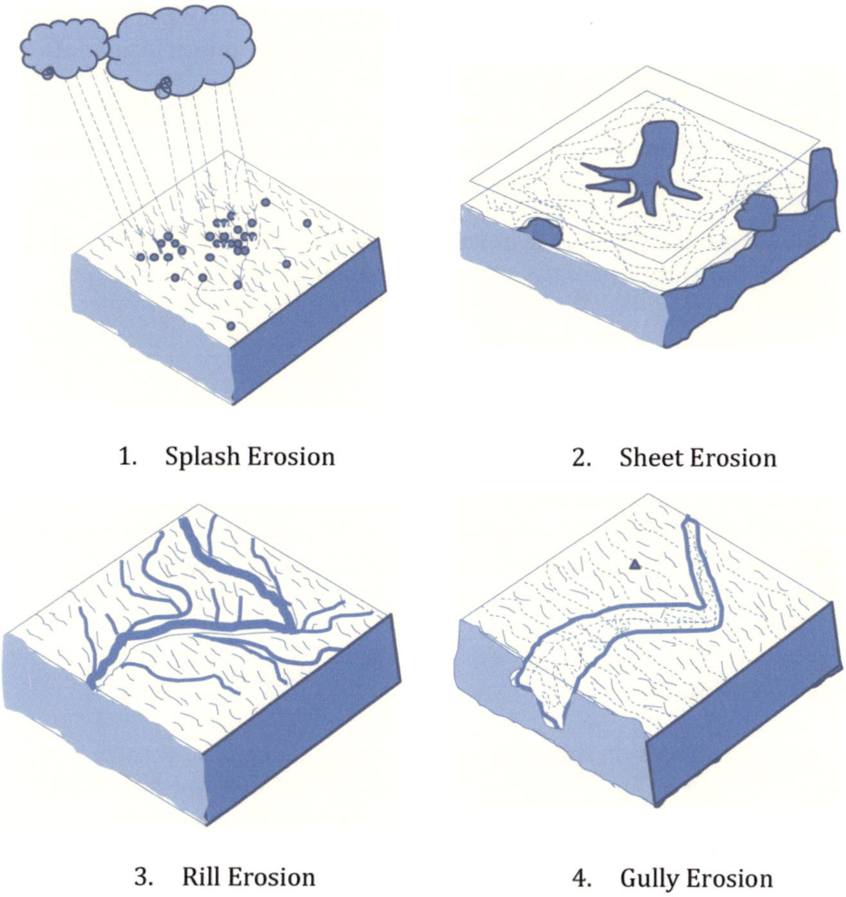

Figure 1.
Four basic forms of soil erosion by water.

4.3 Rill erosion

It describes the flow of runoff water loaded with soil particles and organic matter in finger-like small channels, known as rill erosion. This is the advanced form of sheet erosion for soil loss. Water flow in small channels erodes soil at a faster rate than sheet erosion. Rill erosion is the second most common form of water erosion. These rills can be easily managed by tillage operations but can cause higher soil loss during intensive rainfall. The key factors that cause rill erosion are soil erodibility, land slope, runoff transport capacity, and hydraulic shear of water flow.

4.4 Gully erosion

Gully erosion is the advanced form of rill erosion. When the volume and velocity of concentrated runoff water increase, the rills become deep and broad and forms gullies. The gullies are linear incision channels with 0.3 m width and 0.3 m depth. Concentrated runoff flow is a primary factor for gully formation. Continuous gully erosion results in the removal of the entire soil profile. The extreme form of gully erosion may results in failure of crops, expose plant roots, reduce the groundwater level, and adversely affects landscape stability. It can cut apart the fields and aggravate the non-point source pollution (e.g., sediment, chemicals) to nearby water bodies. Gullies cannot be corrected by usual tillage operations. The dominant factors affecting gully erosion are shear stress of flowing water and critical shear stress of the soil. The further erosion of gullies results in ravines formation. Based on the size, depth, and drainage area, gullies can be classified as:

a. **U-shaped gullies:** These types of gullies are usually formed in alluvial soils where the characteristics of both the surface and subsurface soils are similar.

b. **V-shaped gullies:** This is the most common shape of gully erosion which occurs in the areas where the subsurface of soil is more resistant than the topsoil surface.

4.5 Ravine formation

It is referred to as a network of deep and narrow gullies that flows parallel to each other while linking with the river system. Mismanagement and non-judicious use of land result in enlargement of rills and gullies and eventually lead to ravine formation. Abrupt changes in elevation of the river bed and the adjoining land surface, deep and permeable soil with high erodibility, sparse vegetation, and backflow of river water during the recession period causes severe bank erosion which consequently results in ravine formation.

4.6 Tunnel erosion

It is the sub-soil erosion through runoff flow in channels while surface soil remains intact. Tunnel erosion is also known as pipe erosion and commonly occurs in arid and semiarid regions where the soil permeability for water varied with the soil profile. The further widening and deepening of tunnels form large gullies which degrade the productive agricultural lands. Soil with erodible characteristics, having sodic B horizon and stable A horizon are highly prone to tunnel erosion. Runoff flow through natural cracks and animal burrows initiates tunnel formation by infiltrating thorough dispersible subsoil layers. Seepage, lateral flow, and interflow are key indicators of tunnel erosion. It alters the geomorphic and hydrologic characteristics of the affected areas. Management practices for tunnel erosion are ripping, contour farming, vegetation including trees and deep-rooted grasses with proper fertilization and liming, consolidation of surface soil, and diversion of concentrated runoff.

4.7 Slip erosion or landslip erosion

It is the downward and outward movement of slope forming materials composed of natural rocks and debris from sloppy lands. It is also known as mudslide or mass erosion. This type of erosion mostly occurs in hilly regions having water-saturated soils slips down the hillside or mountain slope. Banks along highways, streams, and ocean fronts are often subject to landslides. The large masses of land slip down which destroy the vegetation and degrade the productivity of lands. The slope can be stabilized through developments of diversion drains, contour trenches, crib structures, geotextiles, kutta—crate structures, and retaining walls.

4.8 Stream bank erosion

The scouring of soil material from the stream bed and cutting of stream bank by the action of flowing water is known as stream bank erosion. Streams and rivers change their direction of flow by cutting the bed from one side and depositing the sediment to the other side of the stream. Flash floods enhanced the stream bank erosion which is more destructive. Stream and gully erosion are relatively comparable. Primarily, stream bank erosion predominantly occurs at the lower end water tributaries which have a relatively flat slope and continuous flow of water.

4.9 Coastal erosion

Sea level is incessantly rising due which can increase the frequency of occurrence of natural disasters like the tsunami in the coastal areas in the future. Such natural hazards produce strong water waves which can severely erode the seaside areas. It is projected that the erosion rate will be higher in coastal regions in the coming years. The anthropogenic activities leading to coastal erosion are port construction, destruction of mangroves, and beach and river bed mining [23].

4.10 Universal soil loss equation for water erosion

The universal soil loss equation (USLE) was given by Wischmeier and Smith (1978) based on the soil erosion causing factors [24].

$$A = RKL\ SCP \tag{1}$$

where A, mean annual soil loss (metric tons hectare^{-1} year^{-1});
R, rainfall erosivity factor;
K, soil erodibility factor
L, slope-length factor
S, slope-steepness factor;
C, cover and management factor;
P, support practice factor.

Among the above-listed factors, vegetation and to some extent soil can be managed to reduce the rate of the soil erosion but the climatic and topographic factors, except slope length, are not manageable. Primarily, soil loss through erosion is a function of erosivity of raindrops and erodibility of the soil which can be mathematically expressed as follows:

$$\text{Erosion} = f\,(\text{Erosivity, Erodibility}) \tag{2}$$

where Erosivity is the potential of rainfall to cause erosion under given soil type and climatic condition; Erodibility is the vulnerability or susceptibility of the soil to erosion which depends on soil bio-physico-chemical properties, and land use and crop management practice. Sandy soils can be easily detached while well aggregated clayey soils are more resistant to erosion than sandy soils. When clay particles detached they can be easily removed by runoff due to their smaller size. Silt soils are the most erodible type of soil [9].

5. Impact of soil erosion on agriculture

The accelerated soil erosion significantly influences the soil quality, agricultural production and nutritional quality [25]. Higher soil erosion results in the removal of fertile topsoil along with nutrients which leads to reduced agronomic yield, land degradation, and terrain deformation [25–27]. The main causal factors affecting the rate of soil erosion are parent material, soil texture, slope steepness, plant cover, tillage, and climate [13]. According to an estimate of existing soil loss data, the mean annual rate of soil erosion in our country is approximately 16.4 ton ha^{-1} which results in annual total soil loss of 5334 million tons (m t) and nutrient loss of 8.4 m t throughout the country [17]. However, the mean annual permissible limit of soil loss is 12.0 tons ha^{-1}. Out of total eroded soil around 29% is permanently lost to the sea, while 61% is transported by runoff from one place to another and the remaining 10% is directly deposited in reservoirs [21]. Higher nutrient concentration has been recorded in soil samples collected from runoff loads over the soil of agricultural fields [28]. Further, around 45.9 kg C ha^{-1} and 4.3 kg N ha^{-1} were recorded in eroded soil during the month of July [29].

The soil organic matter (SOM) is vital for improving soil bio-physico-chemical properties and contains nearly 95% of N and 25–50% of phosphorus [30]. Higher rate of erosion results in loss of soil and fine organic particles. The soil removed by erosion has 1.5–5 times higher SOM than the soil left behind [31]. The availability of SOM also affects the biological activities and soil biodiversity in a particular agro-ecosystem. Moreover, the intensive and erratic rainfall results in higher soil erosion which leads to reduced infiltration and eventually less water availability to the vegetation. Sharda et al. studied the impact of the harshness of water erosion on agricultural productivity and advocated that water erosion reduced the annual crop production by 13.4 Mt in 2008–2009 at the national level [32]. Thus, the soil loss by water and wind severely affects the productive efficiency of all ecosystems [17, 33, 34]. The comprehensive impacts of erosion on soil and water resources which are liable to reduce agricultural productivity are given in **Figure 2** [21].

The vegetation cover is imperative for moderating surface runoff and water erosion from agricultural lands [35]. The rate of runoff, soil, and nutrient loss is predominantly determined by the type of vegetation, canopy cover, slope gradient, and rainfall characteristics [36]. The higher canopy cover and crop residues mulching on soil surface results in the reduced rate of surface runoff and also reduces the impact of rainfall erosivity and soil erodibility [13, 35, 37]. Vegetation cover reduces the detachment of soil particles along with the protection of soil surface from intensive rainfall. Moreover, it also conserves soil moisture and retains sediment and organic materials [38]. To sustain agricultural productivity, it is imperative to reduce runoff, soil loss, and nutrient loss through water erosion [13].

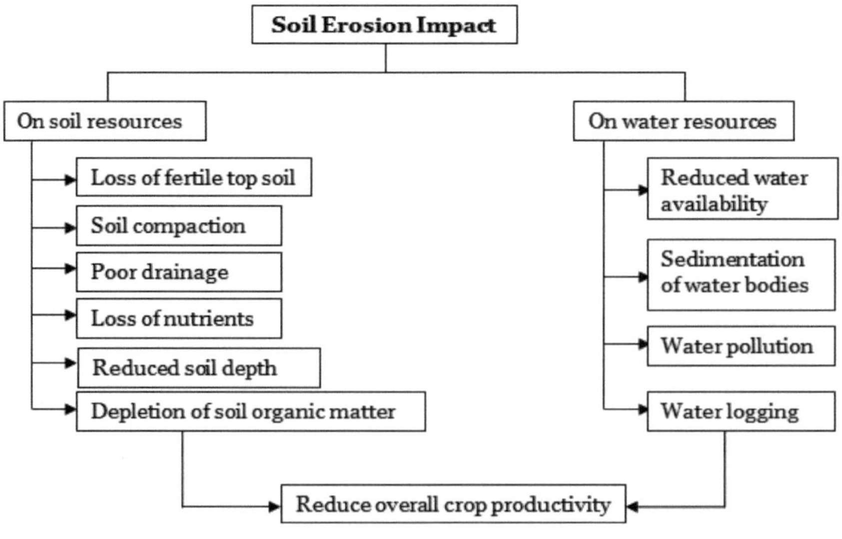

Figure 2.
Impact of erosion on soil and water resources.

6. Soil and water conservation measures

There are two types of measures for soil and water conservation, that is, mechanical/engineering/structural measures and biological measures. Mechanical measures are permanent and semi-permanent structures that involve terracing, bunding, trenching, check dams, gabion structures, loose/stone boulders, crib wall, etc., while biological measures are vegetative measures which involve forestry, agroforestry, horticulture and agricultural/agronomic practices [21].

6.1 Biological measures (agronomic/agricultural and agroforestry)

Agronomic measures are applicable in the landscape of ≤2% slope. Agronomic measures reduce the impact of raindrops through the covering of soil surface and increasing infiltration rate and water absorption capacity of the soil which results in reduced runoff and soil loss through erosion [39]. These measures are cheaper, sustainable, and may be more effective than structural measures, sometimes [4]. Important agronomic measures are described below.

6.1.1 Contour farming

Contour farming is one of the most commonly used agronomic measures for soil and water conservation in hilly agro-ecosystems and sloppy lands. All the agricultural operations viz. plowing, sowing, inter-culture, etc., are practiced along the contour line. The ridges and furrows formed across the slope build a continual series of small barriers to the flowing water which reduces the velocity of runoff and thus reduces soil erosion and nutrient loss [40, 41]. It conserves soil moisture in low rainfall areas due to increased infiltration rate and time of concentration, while in high rainfall areas, it reduces the soil loss. In both situations, it reduces soil erosion, conserves soil fertility and moisture, and thus improves overall crop productivity. However, the effectiveness of this practice depends upon rainfall intensity, soil type, and topography of a particular locality.

6.1.2 Choice of crops

The selection of the right crop is crucial for soil and water conservation. The crop should be selected according to the intensity and critical period of rainfall, market demand, climate, and resources of the farmer. The crop with good biomass, canopy cover, and extensive root system protects the soil from the erosive impact of rainfall and create an obstruction to runoff, and thereby reduce soil and nutrient loss. Row or tall-growing crops such as sorghum, maize, pearl millet, etc. are erosion permitting crops which expose the soil and induce the erosion process. Whereas close growing or erosion resisting crops with dense canopy cover and vigorous root system viz. cowpea, green gram, black gram, groundnut, etc. are the most suitable crops for reducing soil erosion [42]. To increase the crop canopy density, the seed rate should be always on the higher side.

6.1.3 Crop rotation

Crop rotation is the practice of growing different types of crops in succession on the same field to get maximum profit from the least investment without impairing the soil fertility. Monocropping results in exhaustion of soil nutrients and deplete soil fertility. The inclusion of legume crops in crop rotation reduces soil erosion, restores soil fertility, and conserves soil and water [43]. Further, the incorporation of crop residue improves organic matter content, soil health, and reduces water pollution. A suitable rotation with high canopy cover crops helps in sustaining soil fertility; suppresses weed growth, decreases pests and disease infestation, increases input use efficiency, and system productivity while reducing the soil erosion [42].

6.1.4 Cover crops

The close-growing crops having high canopy density are grown for protection of soil against erosion, known as cover crops. Legume crops have good biomass to protect soil than the row crops. The effectiveness of cover crops depends on crop geometry and development of canopy for interception of raindrops which helps in reducing the exposure of soil surface for erosion. It has been reported that legumes provide better cover and better protection to land against runoff and soil loss as compared to cultivated fallow and sorghum. The most effective cover crops are cowpea, green gram, black gram, groundnut, etc.

Advantages

- Protection of soil from the erosive impact of raindrops, runoff, and wind.
- Act as an obstacle in water flow, reduce flow velocity, and thereby reduce runoff and soil loss.
- Increase soil organic matter by residue incorporation and deep root system.
- Improve nutrients availability to the component crop and succeeding crops through biological nitrogen fixation.
- Improve water quality and water holding capacity of the soil.
- Improve soil properties, suppress weed growth, and increase crop productivity.

6.1.5 Intercropping

Cultivation of two or more crops simultaneously in the same field with definite or alternate row pattern is known as intercropping. It may be classified as row, strip, and relay intercropping as per the crops, soil type, topography, and climatic conditions. Intercropping involves both time-based and spatial dimensions. Erosion permitting and resisting crops should be intercropped with each other. The crops should have different rooting patterns. Intercropping provides better coverage on the soil surface, reduces the direct impact of raindrops, and protects soil from erosion [36, 43].

Advantages

- High total biomass production.
- Efficient utilization of soil and water resources.
- Reduction of marketing risks due to the production of a variety of products at different periods.
- Drought conditions can be mitigated through intercropping.

- Reduce the weed population and epidemic attack of insect pests or diseases.
- It improves soil fertility.

6.1.6 Strip cropping

Growing alternate strips of erosion permitting and erosion resistant crops with a deep root system and high canopy density in the same field is known as strip cropping. This practice reduces the runoff velocity and checks erosion processes and nutrients loss from the field [36, 44]. The erosion resisting crops protects soil from beating action of raindrops, reduces runoff velocity, and thereby increased time of concentration which results in a higher volume of soil moisture and increased crop production [4]. Strip cropping is practiced for controlling the run-off and erosion and thereby maintaining soil fertility.

Types of strip cropping

i. **Contour strip cropping:** The growing of alternate strips of erosion permitting and erosion resisting crops across the slopes on the contour is known as contour strip cropping. It reduces the direct beating action of raindrops on the soil surface, length of the slope, runoff flow and increases rainwater absorption into the soil profile.

ii. **Field strip cropping:** In this practice the field crops are grown in more or less parallel strips across fairly uniform slopes, but not on exact contours. It is useful on regular slopes and with soils of high infiltration rates, where contour strip cropping may not be practical.

iii. **Wind strip cropping:** It consists of the planting of tall-growing row crops (such as maize, pearl millet, and sorghum) and close or short growing crops in alternately arranged straight and long, but relatively narrow, parallel strips laid out right across the direction of the prevailing wind, regardless of the contour.

iv. **Permanent or temporary buffer strip cropping:** It is the growing of permanent strips of grasses or legume or a mixture of grass and legume in highly eroded areas or in areas that do not fit into regular rotation, i.e. steep or highly eroded, slopes in fields under contour strip cropping. These strips are not practiced in normal strip cropping and generally planted permanent or temporary basis.

6.1.7 Mulching

Mulch is any organic or non-organic material that is used to cover the soil surface to protect the soil from being eroded away, reduce evaporation, increase infiltration, regulate soil temperature, improve soil structure, and thereby conserve soil moisture [45–47]. Mulching prevents the formation of hard crust after each rain. The use of blade harrows between rows or inter-culture operations creates "dust mulch" on the soil surface by breaking the continuity of capillary tubes of soil moisture and reduces evaporation losses. Mulching also reduces the weed infestation along with the benefits of moisture conservation and soil fertility improvement. Hence, it can be used in high rainfall regions for decreasing soil and water loss, and in low rainfall regions for soil moisture conservation. Organic mulches improve organic matter and consecutively improving the water holding capacity, macro and micro fauna biodiversity, their activity, and fertility of the soil [48, 49].

Inorganic mulches have a longer life span than organic mulches and can reduce soil erosion, water evaporation losses, suppress weeds but cannot improve soil health. This practice is costly and labor intensive therefore, suitable for cash crops such as fruits and vegetables. Polyethylene mulch is commonly used for the conservation of soil and water resources to increase crop productivity [21].

6.1.8 Conservation tillage

In this practice at least 30% of soil surface should remain covered with crop residue before and after planting the next crop to reduce soil erosion and runoff, as well as other benefits such as C sequestration. This term includes reduced tillage, minimum tillage, no-till, direct drill, mulch tillage, stubble-mulch farming, trash farming, strip tillage, etc. The concept of conservation tillage is widely accepted in large scale mechanized crop production systems to reduce the erosive impact of raindrops and to conserve the soil moisture with the maintenance of soil organic carbon. Conservation tillage improves the infiltration rate and reduces runoff and evaporation losses [4]. It also improves soil health, organic matter, soil structure, productivity, soil fertility, and nutrient cycling and reduces soil compaction [50].

6.1.9 Organic farming

Organic farming is an agricultural production system that devoid the use of synthetic fertilizers or pesticides and includes organic sources for plant nutrient supply viz. FYM, compost, vermicompost, green manure, residue mulching, crop rotation, etc. to maintain a healthy and diverse ecosystem for improving soil properties and ensuring a sustained crop production. It is an environmentally friendly agricultural crop production system.

The maintenance of high organic matter content and continuous soil surface cover with cover crops, green manure, and residue mulch reduce the soil erosion in organic farming. It leads to the addition of a large quantity of organic manures which enhances water infiltration through improved bio-physico-chemical properties of soil, and eventually reduces soil erodibility [51]. Organic materials improve soil structure through the development of soil binding agents (e.g., polysaccharides) and stabilizing and strengthening aggregates which reduce the disintegration of soil particles and thus reduced soil erosion. Soil erosion rates from soils under organic farming can be 30–140% lower than those from conventional farming [9].

6.1.10 Land configuration techniques

Adoption of appropriate land configuration and planting techniques according to crops, cropping systems, soil type, topography, rainfall, etc. help in better crop establishment, intercultural operations, reduce runoff, soil and nutrient loss, conserve water, efficient utilization of resources and result in higher productivity and profitability. Ridge and furrow, raised bed and furrow, broad bed and furrow, and ridging the land between the rows are important land configuration techniques.

i. **Ridge and furrow system:** Raising rainy season crops on ridges and *rabi* season crops in furrows reduces the soil crusting and ensures good crop stand over sowing on flat beds. Moreover, inter-row rainwater can be drain out properly during the monsoon period and collected in farm ponds, for life-saving irrigations and profile recharging for the establishment of *rabi* crops. It leads to the increased moisture content in soil profile which reduces moisture stress on plants during the drought period. This method is most suitable for wide-spaced crops viz. cotton, maize, vegetables, etc.

ii. **Broad bed and furrow system:** This system has been developed by the ICRISAT in India. It is primarily advocated for high rainfall areas (>750 mm) having black cotton soils (Vertisols). Beds of 90–120 cm width are formed, separated by sunken furrows of about 50–60 cm wide and 15 cm depth. The preferred slope along the furrow is between 0.4 and 0.8% on Vertisols. Two to four rows of the crop can be grown on the bed, and the width and crop geometry can be adjusted to suit the cultivation and planting equipment.

Advantages

- Increase *in-situ* soil moisture conservation
- Safely dispose of excess runoff without causing erosion
- Improved soil aeration for plant growth and development
- Easier for weeding and mechanical harvesting
- It can accommodate a wide range of crop geometry.

6.1.11 Agroforestry measures

Agroforestry is a sustainable land management system which includes the cultivation of trees or shrubs with agricultural crops and livestock production simultaneously on the same piece of land [52, 53]. It is an emerging technology for effective soil and water conservation and comprises a wide range of practices for controlling soil erosion, developing sustainable agricultural production systems, mitigating environmental pollution, and increasing farm economy. The leaf litter addition act as a protective layer against soil erosion improves soil health and moisture retention capacity of the soil and increases crop productivity [54–56]. It has been reported that different agroforestry practices can reduce up to 10% of soil erosion [57]. Agroforestry not only controls soil erosion but also produce tree-based several marketable products.

Types of agroforestry systems

Agri-Silviculture: It is the growing of agricultural crops as a primary component with the secondary component of multipurpose trees (MPTs) on the same managed land unit. The tree species bind soil particles in the root zone and increase water infiltration, and reduce runoff.

Agri-Horticulture: Growing of agricultural crops and fruit trees on the same managed land unit is known as agri-horticulture. Fruit tree species like lemon (*Citrus limon*), mango (*Mangifera indica*), ber (*Ziziphus mauritiana*), and aonla (*Phyllanthus emblica*) can be successfully planted in agricultural fields and on degraded and low fertile lands with some restoration measures.

Alley Cropping: Growing of agricultural crops in the alley formed between the hedge rows of leguminous nitrogen-fixing tree species. This system is one of the effective measures for soil and water conservation in hilly areas.

Silvi-pasture System: Raising grasses or livestock with MPTs on the same managed land unit is known as silvi-pasture system. This system has the potential to reclaim eroded and degraded lands. Mechanical measures combined with grass species cultivation are more effective for controlling soil erosion processes [58]. The grass species such as *Cenchrus ciliaris* (buffel grass), *Cenchrus setigerus* (birdwood grass), *Dichanthium annulatum* (marvel grass), *Panicum antidotale* (blue panicgrass), *Panicum maximum* (Guinea grass), *Brachiaria mutica* (para grass) and *Pennisetum purpureum* (elephant grass) are important in ravine restoration [59].

6.2 Mechanical measures

Mechanical measures or engineering structures are designed to modify the land slope, to convey runoff water safely to the waterways, to reduce sedimentation and runoff velocity, and to improve water quality. These measures are either used alone or integrated with biological measures to improve the performance and sustainability of the control measures. In highly eroded and sloppy landscape biological measures should be supplemented by mechanical structures. A number of permanent and temporary mechanical measures are available such as terraces, contour bunding, check dams, gabions, diversion drains, geo-textiles, etc. [43]. The mechanical measures are preferred based on the severity of erosion, soil type, topography, and climate [4].

6.2.1 Bunding

i. **Contour bunding:** Contour bunding is used to conserve soil moisture and reduce erosion in the areas having 2–6% slope and mean annual precipitation of <600 mm with permeable soils [60]. The vertical interval between two bunds is known as the spacing of bunds. The spacing of bund is dependent on the erosive velocity of runoff, length of the slope, slope steepness, rainfall intensity, type of crops, and conservation practices.

ii. **Graded bunding:** Graded bunds are made to draining out of excess runoff water safely in areas having 6–10% land slope and receiving rainfall of >750 mm with the soils having infiltration rate < 8 mm/h.

iii. **Peripheral bunds:** Peripheral bunds are constructed around the gully head to check the entry of runoff into the gully. It protects the gully head from being eroded away through erosion processes. It creates a favorable condition for the execution of vegetative measures on gully heads, slopes, and beds.

6.2.2 Contour trenching

Trenches are constructed at the contour line to reduce the runoff velocity for soil moisture conservation in the areas having <30% slope. Bunds are formed on the downstream side of trenches for the conservation of rainwater. Trenches are of two types:

i. **Continuous contour trenches:** Continuous contour trenches are constructed based on the size of the field in the low rainfall areas with the 10–20 cm trench length and 20–25 cm equalizer width without any discontinuity in trench length (10–20 m).

ii. **Staggered contour trenches (STCs)** Generally, these trenches are constructed in alternate rows directly beneath one another in a staggered manner in the high rainfall areas, where the risk of overflow is prominent. SCTs are 2–3 m long with 3–5 m spacing between the rows. Planting of tree species is done based on the land slope. It is highly effective in forestalling extension of gully head, soil loss, and arrest the overflow.

6.2.3 Terracing

Terraces are earthen embankments built across the dominant slope partitioning the field in uniform and parallel segments [9]. Generally, these structures are combined with channels to convey runoff into the main outlet at reduced velocities. It reduces the degree and length of slope and thus reduced runoff velocity, soil erosion and improves water infiltration [5]. It is recommended for the lands having a slope of up to 33%, but can be adopted for lands having up to 50–60% slope, based on socio-economic conditions of a particular region. Where plenty of good-quality stones are available, stone bench terracing is recommended. Sometimes, semi-circular type terraces are built at the downstream side of the plants, known as half-moon terraces. Based on the slope of benches, the bench terraces are classified into the following categories:

i. **Bench terraces sloping outward:** These types of terraces are used in low rainfall areas having permeable soils. A shoulder bund is provided for stability of the edge of the terrace and thus has more time for rainwater soaking into the soil.

ii. **Bench terraces sloping inward (hill-type terraces):** These types of bench terraces are suitable for heavy rainfall areas where a higher portion of rainfall is to be drained as runoff. For this, a suitable drain should be provided at the inward end of each terrace to drain the runoff. These are also known as hill-type terraces.

iii. **Bench terraces with level top:** These types of terraces are suitable for uniformly distributed medium rainfall areas having deep and highly permeable soils. These are also known as irrigated bench terraces because of their use in irrigated areas.

6.2.4 Contour wattling

Wattling is a technique of dividing the length of the slope into shorter sections and in these sections, the wattles are constructed at a vertical interval of 5–7 m up to 33% slope and 3 m up to 66% slope. It is not effective on slopes steeper than 66% and on very loose or powdery rocks [61].

6.2.5 Crib structures

Crib structures are used to stabilize the steep slopes of >40% by constructing log wood structures filled with stone/brushwood. Eucalyptus poles with 2–3 m length and 8–12 cm diameter can be used for the construction of crib structures. These poles are joined together with the help of 20–25 cm long nails. The height of the structure is kept 1.5–2 m above the ground depending upon the land slope [62].

6.2.6 Geo-textiles

Geo-textiles are made up of natural fibers of jute or coir, which are used for stabilization of degraded slopes in mine spoil and landslides areas along roadsides. It facilitates the initial establishment of vegetation on highly degraded sloping lands by holding the vegetation in place and conserving moisture. The open mesh size of geo-textiles varies from 3 to 25 mm. The biodegradability of geo-textiles was reported for 2–3 years. It can absorb 12–25% water under 65 and 95% humidity, respectively and when fully soaked in water it can absorb 40% moisture [63].

6.2.7 Loose boulder/stone/masonry check dams

Check dams are effective for preventing runoff rate and severe erosion in steep and broad gullies, and most suitable for high elevation areas of the catchment [62]. These structures are cheap, having a long life, and fewer maintenance requirements. The depth of gully bed is kept about o.3 m and flat stones of 20–30 cm size are used for the construction of dams. A spillway is provided in the middle of the dam to allow the safe discharge of runoff water [21, 60]. Similarly, gabion check dams are also used for drainage line treatment in sharp slanted gullied areas to check sedimentation, erosion, and to conserve soil moisture [62].

6.2.8 Brushwood check dams

Branches of tree and shrub species are staked in two rows parallel to each other filled with brushwood and laid across the gully or way of the flow. These are usually built to regulate the overflow in small and medium gullies which are supplemented with vegetative barriers for long term effectiveness. There is enough soil volume to establish the vegetation. The tree species are planted in 0.3 m × 0.2 m trenches across the way of gullies. It reduces the runoff velocity, soil loss, and improves soil moisture which helps in the successful establishment of vegetative barriers.

6.2.9 Diversion drains

The channels are constructed to protect the downstream area and for safe draining and diverting of runoff water. It is applicable in high rainfall areas to control runoff losses during the initial stage. The gradient of diversion drain should preferably be kept within 0.5%. Generally, a narrow and deep drain does not get silted up as rapidly as a broad and shallow drain of the same cross-sectional area. Soil dug from the drain should be dumped on the lower side of the drain. Outlet end should be opened at natural drainage lines.

6.2.10 Conservation bench terrace

In the conservation bench terrace (CBT) system, the land is divided into 2:1 ratio along the slope in which the upper 2/3 area (Donor area) contributes runoff to the lower 1/3 runoff collecting area (recipient area). The donor area is left in its natural slope condition. It is also known as the zingg terrace and developed by Zingg and Hauser in 1959. The runoff contributing area is used for cultivation of *kharif* while the lower 1/3 area with conserved soil moisture is used to cultivate *rabi* crops. This mechanical measure can be successfully applied in a semi-arid climate on mild sloppy lands (2–5%) for erosion control, water conservation, and improvement of crop productivity. This system can be used in silty loam to silty clay loam soils. CBT system resulted in the reduction of runoff from 36.3 to 7.4% and soil loss from 10.1 to 1.19 Mg ha^{-1} as compared to the conventional system of sloping border [64]. An average reduction of 78.9 and 88.0% in runoff and soil loss, respectively reported in the CBT system over the conventional system [65].

7. Conclusion

The land is finite and diminishing gradually due to the increasing rate of varied kinds of degradation and thus there is no alternative to expend cultivable land area.

The only way is either increasing agricultural productivity per unit resource available or restoring the degraded lands. Healthy soil and availability of water are vital for productivity in all kinds of terrestrial ecosystems because plants require fertile soil with improved bio-physico-chemical properties and good quality of water for their growth and development. Use of soil and water conservation measures including biological (agroforestry and agricultural) and mechanical measures (terracing, bunding, trenching, check dams, etc.) is imperative to reduce runoff, soil erosion and to improve soil quality, water quality, moisture conservation, and overall crop productivity in a sustainable way. Biological measures are economically feasible and environmental friendly; also improve soil properties along with the conservation of soil and water resources. Further, the combined use of biological and mechanical measures will help in improving and sustaining agricultural productivity.

8. Future perspectives for soil and water conservation

The burgeoning world population, food insecurity and natural resource degradation are the major issues in the present era of climate change. It has been projected that the world population will be ~10 billion in 2050 [66]. Further, the rapid industrial growth and intensive farming practices are expected to increase the pressure on land and water resources in near future. Therefore, a paradigm shift in soil and water conservation, and its management is needed for agricultural sustainability. The some of the future concern for soil and water conservation and sustainable agriculture are the following:

- Formulation of new policies and development of new technologies based on social, economical and cultural aspect of a particular regional.
- Implementation and adoption of effective conservation measures for sustaining agricultural productivity.
- Existing soil and water conservation practices should be improved and developed based on the level of natural resources degradation.
- Greater emphasis should be given on participatory approach for effective soil and water conservation.
- Post impact assessment and monitoring of soil and water conservation measures should be done to evaluate their efficacy in increasing productivity, monetary returns, and livelihood of the stakeholders.
- Development of cost effective conservation practices to restore the degraded lands and to sustain agricultural productivity.
- The efficient technologies for soil and water conservation should be demonstrated on farmers' fields with their active participation.
- Emphasis on research, education and extension of soil and water conservation effective technologies to the stakeholders.
- Adoption of efficient management practices and judicious use of soil and water resources.

Abbreviations

C	carbon
CAZRI	Central Arid Zone Research Institute
CSSRI	Central Soil Salinity Research Institute
ICAR	Indian Council of Agricultural Research
ICRISAT	International Crops Research Institute for the Semi-arid Tropics
IISWC	Indian Institute of Soil and Water Conservation
N	nitrogen
NBSS&LUP	National Bureau of Soil Survey and Land Use Planning
NRSA	National Remote Sensing Agency

Author details

Anita Kumawat[1]*, Devideen Yadav[2], Kala Samadharmam[1] and Ittyamkandath Rashmi[1]

1 ICAR-Indian Institute of Soil and Water Conservation, Research Centre, Kota, Rajasthan, India

2 ICAR-Indian Institute of Soil and Water Conservation, Dehradun, India

*Address all correspondence to: akumawat333@gmail.com

References

[1] Eswaran H, Lal R, Reich PF. Land degradation: An overview. In: Bridges EM, Hannam ID, Oldeman LR, et al., editors. Responses to Land Degradation. Proceedings of the 2nd International Conference on Land Degradation and Desertification, Khon Kaen, Thailand. New Delhi: Oxford Press; 2001

[2] Manivannan S, Thilagam VK, Khola OP. Soil and water conservation in India: Strategies and research challenges. Journal of Soil and Water Conservation. 2017;**16**(4):312-319

[3] UNCCD. Integration of the sustainable development goals and targets into the implementation of the United Nations convention to combat desertification and the intergovernmental working group report on land degradation neutrality. Decision 3/ COP.12. Report of the Conference of the Parties on Its Twelfth Session, Held in Ankara from 12 to 23 October 2015; 2015

[4] Yousuf A, Singh M. Watershed Hydrology, Management and Modeling. Florida: CRC Press; 2019

[5] Gachene CK, Nyawade SO, Karanja NN. Soil and water conservation: An overview. In: Zero Hunger. Encyclopedia of the UN Sustainable Development Goals. Cham: Springer; 2019

[6] Quinton JN, Govers G, van Oost K, Bardgett RD. The impact of agricultural soil Erosion on biogeochemical cycling. Nature Geoscience. 2010;**3**:311-314

[7] Sidhu HS, Jat ML, Singh Y, Sidhu RK, Gupta N, Singh P, et al. Sub-surface drip fertigation with conservation agriculture in a rice-wheat system: A breakthrough for addressing water and nitrogen use efficiency. Agricultural Water Management. 2019;**216**:273-283

[8] Fischer G, Tubiello FN, van Velthaizen H, Wiberg DA. Climate change impacts on irrigation water requirements: Effects of mitigation, 1990-2080. Technological Forcasting & Social Change. 2007;**74**:1083-1107

[9] Blanco H, Lal R. Principles of Soil Conservation and Management. Dordrecht: Springer; 2008. pp. 167-169

[10] Uri ND. Agriculture and the environment—the problem of soil erosion. Journal of Sustainable Agriculture. 2000;**16**:71-94

[11] Pimentel D, Harvey C, Resosudarmo P. Environmental and economic costs of soil erosion and conservation benefits. Science. 1995;**267**:1117-1123

[12] Lal R. Managing soils and ecosystems for mitigating anthropogenic carbon emissions and advancing global food security. Bioscience. 2010;**60**:59-82

[13] Guo M, Zhang T, Li Z, Xu G. Investigation of runoff and sediment yields under different crop and tillage conditions by field artificial rainfall experiments. Water. 2019;**11**(5):1019

[14] Singh RK, Chaudhary RS, Somasundaram J, Sinha NK, Mohanty M, Hati KM, et al. Soil and nutrients losses under different crop covers in vertisols of Central India. Journal of Soils and Sediments. 2020;**20**(2):609-620

[15] NAAS. Degraded and Wastelands of India: Status and Spatial Distribution. New Delhi: Directorate of information and publications of agriculture, Indian Council of Agricultural Research; 2010. pp. 01-158

[16] Mandal D, Sharda VN. Appraisal of soil erosion risk in the eastern Himalayan region of India for soil conservation planning. Land Degradation and Development. 2013;**24**:430-437

[17] Bhattacharyya R, Ghosh BN, Mishra PK, Mandal B, Rao CS, Sarkar D, et al. Soil degradation in India: Challenges and potential solutions. Sustainability. 2015;**7**(4):3528-3570

[18] Indian Council of Agricultural Research (ICAR). In: Trivedi TP, editor. Degraded and Wastelands of India: Status and Spatial Distribution. New Delhi, India: Directorate of Information and Publications of Agriculture; 2010. p. 24

[19] Oldeman LR. The global extent of land degradation. In: Greenland DJ, Szabolcs I, editors. Land Resilience and Sustainable Land Use. Wallingford UK: CAB International; 1994. pp. 99-118

[20] Govers G, Merckx R, Wesemael B, Oost KV. Soil conservation in the 21st century: Why we need smart agricultural intensification. The Soil. 2017;**3**:45-59

[21] Sarvade S, Upadhyay VB, Kumar M, Khan MI. Soil and water conservation techniques for sustainable agriculture. In: Sustainable Agriculture, Forest and Environmental Management. Singapore: Springer; 2019. pp. 133-188

[22] Ghadiri H. Crater formation in soils by raindrop impact. Earth Surface Processes and Landforms. 2004;**29**:77-89

[23] Hegde AV. Coastal erosion and mitigation methods-global state of art. Indian Journa of Geo-Marine Sciences. 2010;**39**(4):521-530

[24] Wischmeier WH, Smith DD. Predicting rainfall erosion losses: a guide to conservation planning. In: USDA, Agriculture Handbook. Washington, DC: U.S. Government Printing Office; 1978. p. 537

[25] Lal R. Soil degradation as a reason for inadequate human nutrition. Food Security. 2009;**1**:45-57

[26] Lal R. Restoring soil quality to mitigate soil degradation. Sustainability. 2015;**7**(5):5875-5895

[27] Wang Y, Fan J, Cao L, Liang Y. Infiltration and runoff generation under various cropping patterns in the red soil region of China. Land Degradation and Development. 2016;**27**:83-91

[28] Young A. Agroforestry for Soil Conservation. Wallingford: CAB International; 1990. p. 318

[29] Santra P, Mertia RS, Kumawat RN, Sinha NK, Mahla HR. Loss of soil carbon and nitrogen through wind erosion in the Indian Thar Desert. Journal of Agricultural Physics. 2013;**13**(1):13-21

[30] Allison FE. Soil Organic Matter and its Role in Crop Production. New York: Elsevier; 1973

[31] Lal R, Stewart BA. Soil Degradation. New York: Springer; 1990

[32] Sharda VN, Dogra P, Prakash C. Assessment of production losses due to water erosion in rain fed areas of India. Journal of Soil and Water Conservation. 2010;**65**(2):79-91

[33] Meena H, Meena RS. Assessment of sowing environments and bio-regulators as adaptation choice for clusterbean productivity in response to current climatic scenario. Bangladesh Journal of Botany. 2017;**46**(1):241-244

[34] Lal M, Mishra SK. Characterization of surface runoff, soil erosion, nutrient loss and their relationship for agricultural plots in India. Current World Environment: An International Research Journal of Environmental Sciences. 2015;**10**(2):593-601

[35] Liu J, Gao G, Wang S, Jiao L, Wu X, Fu B. The effects of vegetation on runoff and soil loss: Multidimensional structure analysis and scale

characteristics. Journal of Geographical Sciences. 2018;**28**(1):59-78

[36] Singh AK, Kumar AK, Katiyar VS, Singh KD, Singh US. Soil and water conservation measures in semi-arid regions of South-Eastern Rajasthan. Indian Journal of Soil Conservation. 1997;**25**(3):186-189

[37] Samra JS, Sharma UC. Soil erosion and conservation. In: Sekhon GS, Chhonkar PK, Das DK, Goswami NN, Narayanasamy G, Poonia SR, Rattan RK, Sehgal JK, editors. Fundamental of Soil Science. New Delhi: Indian Society Soil Science; 2002. pp. 159-170

[38] Kiepe P. No runoff, no soil: Soil and water conservation in hedgerow barrier systems. Tropical Resource Management Papers No. 10; Wageningen; 1995. p. 158

[39] FAO. Guidelines: Land Evaluation for Rainfed Agriculture. Rome: FAO Soils Bulletin; 1984. p. 52

[40] Dimelu MU, Ogbonna SE, Enwelu IA. Soil conservation practices among arable crop farmers In Enugu–north agricultural zone, Nigeria: Implications for climate change. Journal of Agricultural Extension. 2013;**17**(1):184-196

[41] Liu QJ, Zhang HY, An J, Wu YZ. Soil erosion processes on row sideslopes within contour ridging systems. Catena. 2014;**115**:11-18

[42] Sahoo DC, Madhu M, Adhikary PP, Dash CJ, Sahu SS, Devi S. Adoption behaviour of different soil and water conservation measures among tribal farmers of Gajapati, Odisha. Indian Journal of Soil Conservation. 2017;**45**(1):112-116

[43] Vanwalleghem T. Soil erosion and conservation. International Encyclopedia of Geography: People, the Earth, Environment and Technology: People, the Earth, Environment and Technology. 2016;**12**:1-10

[44] Morgan RPC. Soil Erosion and Conservation. Vol. 3. England: Blackwell Science Limited; 2005. p. 315

[45] Pang HC, Li YY, Yang JS, Liang YS. Effect of brackish water irrigation and straw mulching on soil salinity and crop yields under monsoonal climatic conditions. Agricultural Water Management. 2010;**97**:1971-1977

[46] Kumar S, Meena RS, Yadav GS, Pandey A. Response of sesame (*Sesamum indicum* L.) to Sulphur and lime application under soil acidity. International Journal of Plant & Soil Science. 2017;**14**(4):1-9

[47] Jabran K. Role of Mulching in Pest Management and Agricultural Sustainability. Cham: Springer; 2019

[48] Gill HAK, Robert MCS. Effect of integrating soil solarisation and organic mulching on the soil surface insect community. Florida Entomologist. 2010;**93**(2):308-309

[49] Prats SA, Wagenbrenner JW, Martins MA, Malvar MC, Keizer JJ. Hydrologic implications of post-fire mulching across different spatial scales. Land Degradation and Development. 2016;**27**(5):1440-1452

[50] Sharma P, Meena RS, Kumar S, Gurjar DS, Yadav GS, Kumar S. Growth, yield and quality of cluster bean (*Cyamopsis tetragonoloba*) as influenced by integrated nutrient management under alley cropping system. Indian Journal of Agricultural Sciences. 2019;**89**(11):1876-1880

[51] Mader P, Flieβbach A, Dubois D. Soil fertility and biodiversity in organic farming. Science. 2002;**296**:1694-1697

[52] Jhariya MK, Surendra S, Bargali A, Raj S. Possibilities and perspectives of agroforestry in Chhattisgarh. In: Zlatic M, editor. Precious

Forests-Precious Earth. Croatia: IntechOpen; 2015. pp. 237-257. DOI: 10.5772/60841

[53] Singh NR, Jhariya MK. Agroforestry and agrihorticulture for higher income and resource conservation. In: Narain S, Rawat SK, editors. Innovative Technology for Sustainable Agriculture Development. New Delhi: Biotech Books; 2016. pp. 125-145

[54] Sharma R, Xu J, Sharma G. Traditional agroforestry in the eastern Himalayan region: Land management system supporting ecosystem services. Tropical Ecology. 2007;**48**(2):189-200

[55] Meena RS, Kumar S, Datta R, Lal R, Vijayakumar V, Britnicky M, et al. Impact of agrochemicals on soil microbiota and management: A review. Land. 2020;**9**:34. DOI: 10.1016/j.geoderma.2019.114164

[56] Gupta B, Sarvade S, Mahmoud A. Effects of selective tree species on phytosociology and production of understorey vegetation in mid-Himalayan region of Himachal Pradesh. Range Management and Agroforestry. 2015;**36**(2):156-163

[57] Udawatta RP, Krstansky JJ, Henderson GS, Garrett HE. Agroforestry practices, runoff, and nutrient loss: A paired watershed comparison. Journal of Environmental Quality. 2002;**31**:1214-1225

[58] Meena RS, Kumar V, Yadav GS, Mitran T. Response and interaction of *Bradyrhizobium japonicum* and *Arbuscular mycorrhizal* fungi in the soybean rhizosphere: A review. Plant Growth Regulators. 2018;**84**:207-223

[59] Gupta SK, Raina NS, Sehgal S. Potential of silvi-pastoral systems in improving the forage production in the hills of Jammu and Kashmir. Journal of Research SKUAST-J. 2007;**6**(2):160-168

[60] Shinde R, Sarkar PK, Thombare N, Naik SK. Soil conservation: Today's need for sustainable development. Agriculture & Food: e-Newsletter. 2019;**1**(5):175-183

[61] Singh A, Verma SK. Management of ravines through anicuts and afforestation. In: Ravine Lands: Greening for Livelihood and Environmental Security. Singapore: Springer; 2018. pp. 477-504

[62] Meena RS, Lal R, Yadav GS. Long term impacts of topsoil depth and amendments on soil physical and hydrological properties of an Alfisol in Central Ohio. USA. Geoderma. 2020;**363**:1141164

[63] Vishnudas S, Anil KR, Savenije HH, Van Der Zaag P. Use of coir geotextiles for soil and water conservation: Case Studies from India. In: Technological Interventions in Management of Irrigated Agriculture. Canada: Apple Academic Press; 2018. pp. 161-209

[64] Singh A, Meena RS. Response of bioregulators and irrigation on plant height of Indian mustard (*Brassica juncea* L.). Journal of Oilseed Brassica. 2020;**11**(1):9-14

[65] Meena H, Meena RS, Rajput BS, Kumar S. Response of bio-regulators to morphology and yield of clusterbean [*Cyamopsis tetragonoloba* (L.) Taub.] under different sowing environments. Journal of Applied and Natural Science. 2016;**8**:715-718

[66] Nations U. World Population Prospects: The 2017 Revision, Key Findings and Advance Tables. New York: United Nations; 2017

Chapter 3

Forest Soil Water in Landscape Context

Aleš Kučera, Pavel Samec, Aleš Bajer, Keith Ronald Skene, Tomáš Vichta, Valerie Vranová, Ram Swaroop Meena and Rahul Datta

Abstract

Forests play an irreplaceable role in linking the water cycle with the functions of soil. Soil water not only enhances the stability of forests, but also its run-off and evaporation affects the growth of plants in different ecosystems. The forest soil water balance is contextualized within the immediate and more global landscapes, in terms of relations of water to the soil environment and bedrock, participation in the local water cycle within a catchment basin and in the global cycle between ecosystems. Modifications by human civilization can have significant impacts, including erosion intensification, eutrophication, salinization, spreading of single-species plantations, and regime shifts. Forests regulate the movement of water in the soil environment by reducing the intensity of run-off. Such moderated run-off prevents the occurrence of flash floods, maintaining continuous availability of water for plant and human use. Participation of soil water in the cycling of elements in forests is modified by soil organic matter balance. The preservation of hydric functions in forest soils depends on prioritization of water balance restoration in every catchment basin enclosing the local element cycle. More fundamentally, the development of a synergistically interlinked system, centered around the soil-forest-water-civilization nexus, must become an urgent priority.

Keywords: water potential, available water capacity, forest soil hydric potential, soil water communication, soil water and human society

1. Introduction

In this book section, we deal with four mutually coherent sub-sections which, according to the author teams should present the topic progressively from base soil-water interactions, properties and parameters on general level (Section 2.1); landscape and forest-horizontal water relations (Section 2.2); landscape and forest-vertical water relations (Section 2.3) and holistic soil-water-forest-landscape-civilization nexus (Section 2.4).

Soil water refers to any water contained in the soil in liquid, gaseous and solid states. From a forestry point of view, water can be considered as a key factor of production and its sustainability, while also contributing to the stability of the forest ecosystem, since water is essential, not only for nutrition (both as a reaction medium and as a substrate), but also for the growth and development of stands. Soil water in the liquid state acts by its deflocculating, dissolving, hydrolytic and translocation effects. Soil water is irreplaceable in a wide range of Physico-chemical, biochemical and biological processes and de facto it conditions soil formation and the development of the pedosphere. Oxygen, upon which all anaerobic life depends, is generated from the water-splitting reaction. Entire photosynthetic physiological pathways, such as Crassulacean Acid Metabolism (CAM), are engineered around water conservation.

2. Soil water and its relation to the soil environment

2.1 General characteristics of soil water

Water exists as a soil solution in the soil [1]. Gases (O_2, CO_2, NH_3, N-oxides, S-oxides, etc.) and minerals are dissolved in this solution. Dissolved mineral substances originate from weathering processes, where they are released from rocks into the soil solution, and also from the above-ground part of forest stands, either by means of emission or percolation through tree crowns. Up to 50–250 kg of minerals per hectare a year penetrate the soil by so-called 'cloud/fog water' [2, 3]. This results in a significant enrichment of the soil surface not only in the form of plant litter but also through rainwater, including such elements as Ca, Mg, K, P and N. These elements react in the solid phase in the soil, further dissolving or precipitating. The water composition depends on the dissolution of minerals and organic compounds, on the ion exchange between the soil sorption complex and the soil solution and on the interaction of the soil solution, fine roots and soil microorganisms. Mineral (acids, bases, salts) and organic substances (colloids of dissolved compounds, saccharides, fulvic acids and amino acids, expressed as dissolved organic carbon (DOC)) are dissolved in water and then pass through the biosphere, while being regulated by climatic factors. Due to climate change and associated substantial changes in forest stand structure and functioning, the cycles and flows also change, not only at the level of soil water percolation and content but also within bulk deposition and through fall, both representing substantial sources of DOC [4].

2.1.1 Water sources and losses in forest soils

The soil water content and its availability are the results of a water balance arising from the inputs and outputs of the water cycle within the particular ecosystem [5, 6]. The actual soil moisture enters and leaves the water balance at the beginning and the end of the investigation period respectively. Individual components of the water balance [7] are subject to external influences (generally climate and topography) and internal influences (including properties, composition of the soil body and vegetation characteristics).

The most important water source is vertical precipitation in most areas of the temperate climatic zone. Horizontal precipitation is also regarded as a significant source, for example cloud/fog water in misty forests of tropical or mountain areas, dew, interception, condensation of water vapor in soil pores (especially in soils with a high proportion of macropores), capillary lift and lateral water. The water loss from the soil is primarily due to infiltration, surface run-off and evapotranspiration. Run-off is significantly regulated by forest stands, both in a precipitation-rich period (run-off is lower compared to the non-forested soil) and in a drought (run-off is higher in comparison with the non-forested soil).

The character of surface run-off and water flow through the soil depends on many factors, notably the slope gradient, the amount and intensity of precipitation, soil permeability, the depth of freezing and vegetation coverage. An excessively dried soil surface may be characterized by poor wettability, while humus acts like a permeable filter with high hydraulic conductivity after being soaked in water. This leads to less vulnerability in forest ecosystems compared with different vegetation types [8, 9]. Humus may also be characterized by lower water loss (higher retention) compared to mineral soil. The forest floor, which is typical for forest soils [10], plays a crucial and indispensable role in terms of nutrient supply [11] but also for the water regime [12]: it absorbs several times more water than mineral horizons located below and, at the same time, it reduces soil water losses.

The forestry-pedology nexus represents perhaps the greatest existential threat to humanity at present, requiring urgent action yet currently being ignored by the international community. Historical precedent is all too clear, yet we ignore this growing crisis at our peril. Deficiency of physiological water [13–15] and the potential risk of stress associated with water unavailability to plants [16, 17], both of which differ between vegetation types [18], cannot be overemphasized [19, 20]. The internal factors impacting water availability in the soil environment include the grain-size composition of the soil (the distinction of stoniness and fine earth in a differentiated way in sand, silt and clay fractions), the organic matter form and content and the thickness of soil horizons, affecting both the multidirectional water flow and the physiological depth of root distribution. Other factors include soil chemistry (increased hygroscopicity of salinated soils) [21], the degree of rooting (water drainage alongside the roots) [22] and the distribution and representation of soil pores of specific sizes, but also anthropogenic impacts, such as pedocompaction.

2.1.2 Soil water content, forms and water regime

One measure of increase and loss of water in the soil is the instantaneous soil moisture, represented by the total sum of water sources and losses and the water retention capacity. It is expressed in percentages by volume (Θ) or the mass (w) water content and also mm of water supplied, depending on different applications. In particular, forest soil humus horizons, act differently depending on stand species composition [23], the indicator of volumetric water content is more appropriate than the mass water content. The reason (also associated with low humus bulk density) is a significant disproportion in volumetric and mass water content when the maximum volumetric water content is always less than 100% while the maximum mass water content can be far more than 100% (even exceeding 1000%).

Water is bound to the soil by the range of forces [24, 25] (chemical, physicochemical, physical and biological). The components that, together, produce water potential (see below) act simultaneously to influence water behavior and water content in the soil. There is no sharp boundary between these different forces. As a rule, the water-binding forces in the soil overlap and they are frequently related to specific soil horizons. The resultant sums of forces that hold water in the soil (matrix, osmotic, sorption, capillary, pneumatic, gravitational forces) together make up the soil water potential (Ψ) representing the strength by which soil water is bound. It can be said that it represents energy (work) that we would have to expend to 'drain out' water from the soil. The negative pressure is then referred to as suction or tension; hence it is expressed as the negative value of the atmospheric pressure [−Pa, −kPa, −MPa], where 0.1 MPa = 1 bar = 1020 cm of the water column = 760 mm Hg = 1 atm.

The soil water potential can also be formulated in pF curves, where pF = $-\log \Psi$. The pF curves thus express the relationship between the soil water content and the soil water potential (**Figure 1**).

Water flow in the soil is conditioned by means of two processes [27, 28]: infiltration (determined by field or laboratory infiltration tests), where empty pores are filled with soaked water, and unsaturated flow. This sort of flow gradually slows down until all the pores are filled with water and water flows freely through non-capillary pores. Thus, the soil is fully saturated with water, and saturated flow is realized. This is not uniform, but, rather, tongue-like in terms of the water column, which gradually increases from the soil surface to greater depths.

In sloping landscapes of humid areas, lateral water is also added to rainwater [29]. This means that as we descend a sloping landscape, more water flows on the slope lower down the incline than higher up because soil water from the higher slopes is added to infiltrating rainwater. This phenomenon may also contribute to the differentiation of the soil types over a short distance.

As can be seen from the characteristics of the water potential, water flow in the soil is influenced by moisture gradients, but also by temperature and the mineralogical composition of the soil. The downward direction of water percolation typifies humid areas, where this type of movement contributes to the eluviations of soil particles. Under arid or semi-arid climate conditions, prevailing water flow is upward, as a consequence of suction pressure, and thus water rises by capillary action through the soil profile.

The moisture regime represents the distribution and movement of water in spatial and temporal terms [30]. It incorporates water inputs into the soil, water retention in the soil and water leakage from the soil. The water regime is conditioned by climate, vegetation, the soil-forming substrate, the groundwater location, the terrain relief and the landscape history. The water regime is generally expressed in terms of the relationships among temperature, potential evapotranspiration, precipitation and actual evapotranspiration. The soil water regime can be classified into several categories: aquic, udic, perudic, ustic, aridic and xeric [30]. Based on the resulting balance, there is a water deficit (percolates into underground layers) or a water surplus (retained in the soil).

2.1.3 Soil hydrolimits and plant-available water

Soil hydrolimits (**Figure 1**) represent the strength of water binding in the soil [7, 26, 31, 32]. They denote qualitative and quantitative alterations in soil-water relations, or how strongly water is retained in the soil (in what volume) at the given soil moisture level. Soil hydrolimits are soil moisture values achieved under well-defined conditions and they describe the relation of water and soil according to the flow of water in the soil and its accessibility to plants.

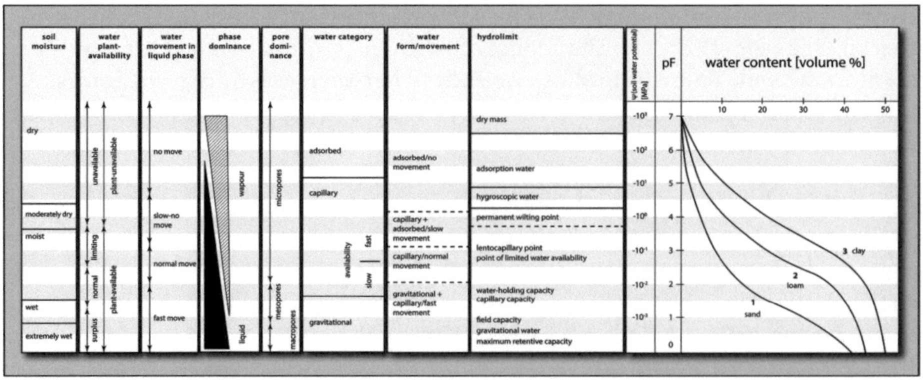

Figure 1.
Relationships between various forms of water and binding forces in the soil (modified according to Vavříček, Kučera [26]).

The significant hydrolimits are:

- Maximum retentive capacity: soil fully saturated with water achieves a hydrolimit, which corresponds to the soil porosity

- Gravitational water: $\Psi = -33$ to -10 kPa or more; under natural conditions, its presence in the soil is qualified mainly by precipitation. The direction of the gravitational water flow is in the direction of gravity to the lower soil strata

- Maximum capillary capacity: volume of capillary and partly semi-capillary pores. Suction forces at this level of the soil water content are in the range of pF 1.6–2.0 ($\Psi = -0.01$ to -0.007 MPa). Only coarse pores are present at this saturation degree without water

- Water-holding capacity (WHC): corresponds to the pF curves in the interval of pF = 2.0–2.7 ($\Psi = -0.08$ to -0.01 MPa), expressing the ability of soil to retain a certain amount of water for a longer period (24 hours). We can ascertain the division of soil pores into capillary and semi-capillary pores by identifying the water-holding capacity

- Point of limited availability: the initial phase of the deteriorated availability of water and its soil mobility. Water still flows continuously through the soil, but merely in the thinnest pores. The water flow is interrupted in semi-capillary and non-capillary pores and water only encapsulates the soil particles

- Lentocapillary point: occurs at pF = 3.0–3.3 ($\Psi = -0.3$ to -0.1 MPa). It is the soil moisture, which is in the range between slightly and scarcely mobile capillary water. It corresponds with a state where a sudden drop in mobility begins by interrupting capillary water

- Wilting point: starts at pF = 4.18 ($\Psi = -2$ to -1 MPa; conventionally -1.5 MPa). It indicates the soil moisture level at which plants are insufficiently supplied with water

- Pellicular water: at pF = 2.1–4.0 (-5.0 to -0.1 MPa). Water encapsulates soil particles in a thicker layer, not moving with gravity, but merely from particles with a larger pellicle to particles with a smaller one. It is unavailable to plants; sometimes it is perceived as hygroscopic water

- Hygroscopic water: is bound to the soil by means of adsorption and osmotic forces. As a rule, it only encapsulates soil particles, and Ψ is generally less than -3.1 MPa, and so it is immobile and unavailable to plants

- Field capacity: represents the ability of the soil to retain the maximum amount of water in the natural profile (in site conditions) against the effects of the Earth's gravity, that is without further active water removal, for a longer period of time (24 hours). This hydrolimit, which is de facto compatible with the water-holding capacity, is widespread, especially in agronomic soil science, where it is also detected by field methods

Plant-available water capacity (PWC) reflects the increase and loss of water in the soil. It is expressed in %, but also in mm of supply [25, 33]. The determination is based on the assumption that a column of water of 1 mm in height represents a water volume of 1 l per 1 m^2. For the practical application of this relation, it is essential that volumetric percentages of the ascertained soil moisture content, or the given hydrolimit, express the soil water supply in mm for a soil layer of 100 mm. In forest soils, this value is depicted in terms of the root distribution for the upper 20 cm of soil, and the observed volumetric % of the soil moisture is therefore multiplied by two to express the value of the plant-available water capacity. The plant-available water capacity formulates the height of the water column of the soil within the range of the wilting point and the water-holding capacity. Thus, PWC represents the condition of the soil moisture where soil water is bound for a relatively long time, but it is still available to plants. The highest values of the plant-available water capacity are in loamy soil. Lower values exist in clay soils, and the lowest values are found in sandy soils [24, 34]. In addition to the texture characteristics, it is necessary to take into account the degree of soil stoniness, which practically does not participate in water retention and represents an inactive soil component in terms of water retention capacity, when determining the plant-available water capacity. PWC also expresses how much torrential rainfall the soil is capable of collecting. From this standpoint, it is an important indication of the water-retaining capacity of the landscape of which the soil is a part as a geological formation, which, with great efficiency, counteracts the flood distribution caused by torrential as well as prolonged rainfall. In this respect, soil, especially forest soils, with several times higher PWC in comparison with agricultural land and much higher than urbanized areas, plays an irreplaceable role in water management in the landscape.

Another soil property, soil moisture storage, relates directly to the actual soil water status, and shares the same units and the same principles as PWC. It can be expressed as the variance between the current soil moisture and the wilting point in mm, representing the current content of physiologically available water.

2.1.4 Soil porosity and capillarity

Apart from the soil structure, porosity is a major factor in the spatial arrangement of the soil and is fundamentally involved in the characterization of water and soil-air regimes, and in the soil–plant (forest stand) relationship. Pores exist in the soil both between soil particles and structural elements (aggregates). If the porosity value between aggregates is marked with the symbol A and the porosity value within the aggregates with the symbol B, the optimum soil porosity may be expressed as A:B = 1:2.

Water is bound most weakly in non-capillary and semi-capillary pores. This kind of water is called gravitational water. Non-capillary porosity occupies pores with very low water retention capacity, in which water moves under the influence of gravity. This is also why the term gravitational water is used for water contained in non-capillary pores. When non-capillary pores dominate, the soil has a low available water content due to its rapid flow to depths unavailable to plant roots.

Capillary water is present only in capillary pores. It is not tied to the Earth's gravity and can move in all directions in the soil. Capillary water is bound thanks to capillary adhesion and the surface tension of menisci. The optimal proportion of capillary pores is approximately two-thirds of total porosity [35–37]. An excess of capillary pores complicates infiltration of water, and it also inflicts an elevation in surface run-off, increasing the risk of erosion. A lack of capillary pores prefigures low plant water supply, low water retention capacity and low water absorption.

Water can rise above a continuous groundwater table by means of capillaries. This is called capillary rise [24, 25, 38]. The capillary rise is approximately the same as the soil particle size (pore diameter = 0.3–0.7 times the soil particle diameter). The capillary rise varies from 10 of centimeters to metres within a given year.

The volume relation of capillary and non-capillary pores is expressed by the minimum aeration capacity [26, 39]. This represents the volume of air-filled pores when the soil has reached maximum capillary capacity. The lower limit value of the minimum aeration capacity of forest soils can be considered to be 8% vol, while the average value (e.g. for topsoil in forest nurseries) is 10% vol. If the soil is excessively aerated, the soil is easier to heat, vapourization increases and soils are contrarily dehydrated. Therefore, a value above 20% can be considered an upper but still acceptable limit, with a risk value of 25%.

2.2 Hydric functions of forest soils

2.2.1 Effect of climate change on forest water cycle

A global (large) water cycle can be defined as a water cycle in which water is transferred between the land and oceans and a local (small) water cycle is defined as a displacement merely over oceans or drainless areas of the land. The water cycle governs all of the natural forest functions. However, forest ecology represents an important aspect of the hydrological cycle at the planetary level, and so these effects impact at a global level. Whereas the global water cycle is related to the adaptation of forests to climate change, the local water cycle interlinks mutual interactions between related forest complexes within the catchment. In general, the impact of forests on global climate change is at its most significant due to cloud formation in the tropics. The formed clouds reflect solar radiation more effectively and, therefore, cool the atmosphere more than does the absorption of greenhouse gases by vegetation [40]. Environmental pollution, deforestation and transformation of the tree species composition reduce the natural ability of forests to adapt to climate change. Monitoring of soil properties focused on water and nutrient cycles in different forest ecosystems offers a tool for assessing the impacts of climate change [41].

Forest functions are the outcome of the interactions between the environmental, soil and vegetation subsystems. Natural functions are based on processes that support self-organization, recovery and development of the ecosystem. The interrelated processes of biodiversity, organic matter formation and nutrient cycles promote production, air circulation, (in-)filtration, evapotranspiration and site differentiation [42]. The water cycle controls the carbon cycle through which forests modify local cycling of all nutrients. The parts of the water and carbon cycles within soils have linked individual forest functions to the self-organized ecosystem [43]. The degree of interconnection is subject to the flow of soil water, but simultaneously also by its scope in the specific cycle.

The global effect of forest functionality consists in the transfer of evaporated water through cloudiness within the catchments from the areas with more significant vapor in the lower parts than the areas in the upper parts, where cloudiness condenses into more frequent precipitation. Precipitation in the upper parts of the catchments flows to the lower parts, where additional water complements the higher evaporation and lower precipitation [44]. As temperatures rise, this phenomenon is intensified: evaporation elevates, and drought deepens in drier areas, while precipitation in wetter areas increases. The consequent accentuation in disparities between drier and wetter areas disrupts the interconnection of forest functions among vegetation zones [45].

Even though the local water cycle defines the hydric functions of forests to catchments, their response to climate change depends on the interconnected monitoring of variability not only within the catchment but also among remote catchments. The link between the effects of global and local water cycles also exposes mutually unrelated forests to reduced water availability and consequently to reduced service provision [46].

2.2.2 Effects of nutrient cycles on forest hydric functions

The forest promotes both water and carbon cycles in parallel because they are related to energy flows in the ecosystem. While natural plant-to-plant feedbacks between plants and nutrient cycles underpin ecosystem functionality, forest damage disrupts these processes. If forest damage results in the disruption of the carbon cycle, at the same time the local water cycle is also disrupted, followed by negative impacts on the functioning of related ecosystems [47]. Recognition of forest function damage through the disruption of soil properties is based on the determination of critical values of physical and chemical properties involved in the processes of formation of individual ecosystem functions.

Carbon enters the ecosystem in the form of atmospheric CO_2 through photosynthesis of plants, in which solar energy for the synthesis of organic compounds in cells is transformed by the decomposition of water. Plants release carbon by respiration, by being consumed by herbivores or fungi and by exchange reactions with soil biota and litter. The most significant conversions of organically bound carbon occur in the soil. Plants mediate carbon into the soil both by litter to the surface and also by root necrosis, exudation, root cap sloughing or exchange with microorganisms within the soil body (e.g. through mycorrhizal sheaths). Litter is mechanically or biochemically decomposed into residual chains at pH $>$ 4.5, or into stable polyphenol nuclei at lower pH, as a result of the tetravalency of carbon covalent bonds. Soil organisms or enzymes are capable of decomposing chains into organic acids under favorable conditions, but the prevailing unfavorable conditions allow merely incomplete decomposition. Soil carbon accumulates as a consequence of the imperfection of decomposition [48]. However, destruents mineralize organic residues back to CO_2 under a range of unfavorable conditions (**Figure 2**).

Carbon compounds significantly attract soil water through adhesion to organic molecular chains. That is why carbon storage in the soil irreplaceably increases the WHC of the entire ecosystem. Subsequently, the detection of forest functions using intra-soil processes focuses on common inputs or outputs of substances and energy in the soil subsystem. This can be done by ascertaining the length of the delay of the forest stand response in the aftermath of the alteration of soil property values [49]. The evaluation of conditions of substance inputs or outputs concentrate on assessments of whether or not biochemical and physical properties can regulate the processes of water or carbon cycles. Even though the selected soil properties correlate with one another, the temporal variability of physical properties is incomparably longer than the significant seasonal variability of biochemical properties. Whereas the variability of (bio)chemical properties indicates a threat to the forest after an episode of drought or extreme daily precipitation sum (EDSP), an alteration to correlations of the forest status with poorly variable soil physical properties indicates deviations in development during environmental change [50].

The selection of intra-soil processes affecting forest functions is based on the study of the variability of properties in different parts of the soil body. The functions of circulation, infiltration, evapotranspiration and differentiation are typically regulated by means of one soil process. The indication of individual forest ecosystem functions at the soil level (**Table 1**) can be derived from the generalization of studies focused on the relation between the growth conditions with water balance, biodiversity and the health status of forests [49, 51–53].

Table 1 shows the soil properties involved in water and carbon cycle processes that increase the efficiency of individual forest functions. The production

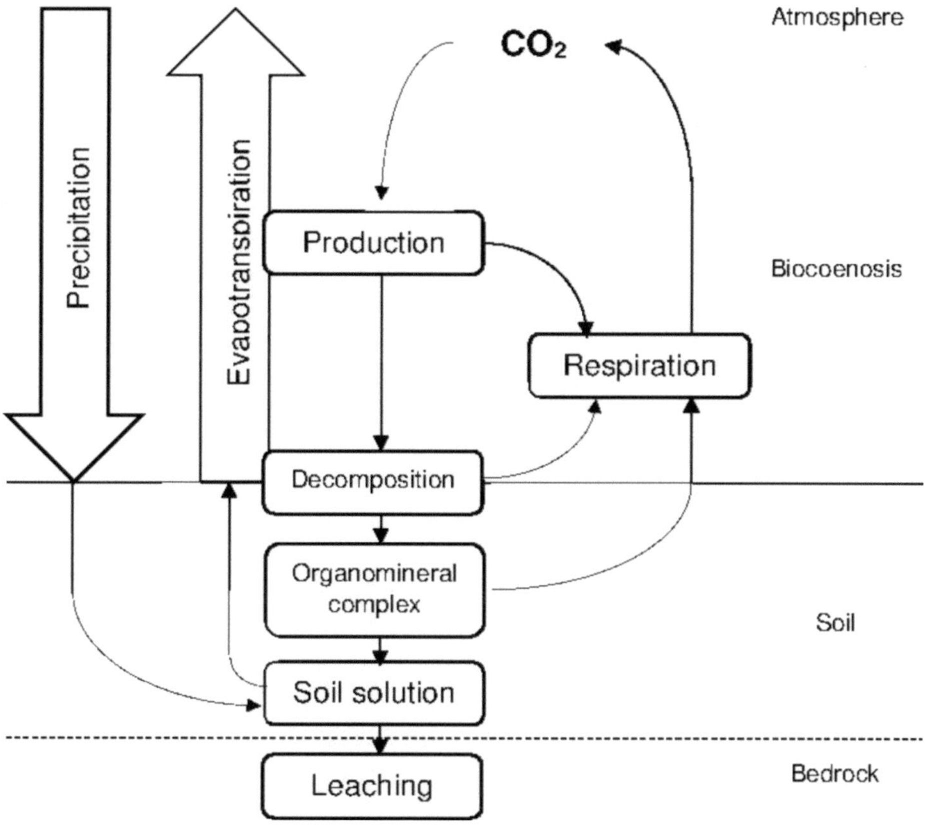

Figure 2.
Connectivity between cycles of water (left) and carbon (right) in forest ecosystem forming hydric functions.

Forest function	Hydric process	Indicative soil property
Production	Photochemical water disintegration	Catalase activity
Air circulation	Vapor pressure decrease	Minimum aeration capacity
(In-)filtration	Physical sorption	Organomineral complex content
Evapotranspiration	Evaporation	Soil water potential
Differentiation	Debasification	Soil water acidity

Table 1.
Relationships between forest functions and water cycle processes indicated by the selected soil properties.

indication is centred on the catalase activity, which depends on the intensity of aerobic metabolism. The correlation of the soil catalase activity with the content and character of organic acids reflects the variety of humus forms. It is naturally associated with differentiated forest cover. If the forest disruption does not damage the humus diversity, the catalase activity remains stable. Air circulation is dependent on the atmospheric flow reducing vapor pressure above the partial surface that the soil maintains thanks to minimum aeration capacity. Infiltration is also conditioned by organic matter and clay minerals, which may form organomineral complexes. They significantly retain water by adhesion and capillary rise in capillary pores remaining among their particles [54]. On the contrary, evapotranspiration is the sum of evaporation from the individual types of surfaces in the ecosystem. The rate of evaporation from the soil is directly proportional to the soil water potential [55].

In contrast, the capability of ecosystem differentiation is estimated by the chemical composition of the soil solution. It grows when soil run-off contains a minimum of base cations. Increased concentrations of bases in run-off water indicate soil acidification, which reduces the ecological diversity of the catchment [56].

2.2.3 Hydrographical division of forests

The soil indicators relating to forest functionality are naturally subdivided into a total of eight biomes relating to differences in water availability due to variations in evapotranspiration and the water-holding capacity (**Table 2**) [57]. Despite the differences between forest biomes, large catchments possess similar zonality of hydric functions internally. Nonetheless, the WHC affects the variances in forest hydric functions of forests between individual habitats within the catchment as its value is directly proportional to the soil types present. The largest overlaps in the WHC values occur in the catchments of Mediterranean, temperate and tropical coniferous forests with more similar soil development at medium temperature intervals relative to boreal, mangrove or tropical broadleaved rainforests [58]. For example, the values of the WHC in **Table 2** were found to be related to the macroclimatic properties of forest biomes. This can be further related to the weighted means of the soil types as found in the Harmonized World Soil Database (HWSD) [59].

Transitions of forest hydric functions in the catchment are the basis for the derivation of hydrographic zonality. Large forest catchments include montane, submontane and floodplain forest ecosystems [50]. These zones emerge thanks to the local water cycle from wetter mountains to drier floodplains. Undisturbed forests are capable of water supply to all the parts of the catchment continuously even though most water supplies on the mainlands are unavailable to plants. Over 62.4% of mainland water supplies are concentrated in glaciers, 36.2% in underground reservoirs and 0.42% in lakes or ground level reservoirs. Only 0.29% of water is found in the soil and 0.09% in rivers [60, 61]. Atmospheric precipitation over the dry land brings merely 0.008% of the global water balance, but over 50% of precipitation occurs in montane areas. Continuous water management in the catchment is ensured by forests by means of modifications of evapotranspiration and run-off. Forests cover 39.7% of the dry land

Forest biome	AP	*T*	*P*	AET	PET	WHC
Tropical rain broadleaved	33.46	21.82 ± 1.35	1988 ± 83	892 ± 200	1270 ± 172	24.96 ± 3.38
Tropical dry broadleaved	5.09	24.16 ± 1.93	1263 ± 79	717 ± 180	1203 ± 174	26.34 ± 5.09
Tropical coniferous	1.20	19.40 ± 2.31	1438 ± 83	716 ± 97	1218 ± 74	33.71 ± 6.97
Temperate mixed	21.71	9.73 ± 7.39	1072 ± 30	508 ± 127	688 ± 112	31.13 ± 9.80
Temperate coniferous	6.91	6.39 ± 7.82	918 ± 36	428 ± 88	700 ± 100	29.63 ± 8.31
Boreal	25.59	−2.54 ± 12.37	642 ± 19	256 ± 75	298 ± 64	49.11 ± 9.15
Mediterranean	5.45	15.05 ± 5.34	586 ± 26	316 ± 121	962 ± 225	32.39 ± 9.44
Mangroves	0.59	26.07 ± 1.70	1900 ± 94	502 ± 476	1303 ± 492	69.18 ± 3.89

AP, area proportion (%); T, average temperature (°C); P, annual precipitation (mm); AET, actual evapotranspiration; PET, potential evapotranspiration; WHC, water-holding capacity (%). Data according to [57].

Table 2.
Characteristics of water balance in forest biomes.

but account for 67.6% of evapotranspiration. Simultaneously, only the structure of the forest can return the evaporated water sufficiently by cloud/fog water or seasonal pollen release, which can create a condensation nucleus to form cloudiness.

Deceleration of run-off by the forest ecosystem is irreplaceable in reducing seasonal variations in water availability between winter and the growing season and in dampening of EDSP. EDSP typically exceeds the average soil WHC either in above-average climatic episodes of precipitation or during the most intense precipitation season. Overcoming WHC prefigures a temporary increase in the flow of soil water and subsequently also river water. It is precise because the values of WHC naturally alter within soil development regardless of the tree species composition or altitude, that (sub)montane forests can dampen run-off after extraordinary rainfall with similar efficiency [62]. The actual water-holding capacity of forest soils due to the constant presence of natural moisture is approximately only 30 mm, providing 22% WHC and dampening 67–75% of EDSP [63].

Alterations in the tree species composition of forests have had the greatest impact on the forest hydric functions during transitions of the seasons of the year. Coniferous trees may be characterized by average higher interception and evapotranspiration. At the same time, coniferous forests capture more snow and significantly slow down melting, reducing the surface run-off in early spring when most of the vegetation is inactive. In deciduous broadleaved forests, this deceleration of run-off does not occur due to defoliation of trees in winter and thus increased solar radiation directly impacting on the soil surface [64].

Hydrographic forest zonality indicates differentiated forest efficiency in the modification of the local water cycle. The differentiation of the effective influence of forests is determined by the relief of the landscape as well as soil development and tree species composition.

Montane forests are located in the upper parts of catchments with the highest amount of precipitation. Their structure is adapted to the application of more frequent horizontal precipitation. Soils are permeable due to the prevailing mechanical weathering. The erosion on steep slopes and the nature of the soil-forming substrate cause rockiness and shallowness of soils. The water-holding capacity of montane soils is maintained by means of accumulation and the slower degradation of humus. Montane drainless depressions with accumulating humus are habitats of ombrogenic bogs in the presence of excessive rainfall. At transitions of the mantle rock with the outcrop of impermeable subsoil, there are water springs at the points of concentrated groundwater run-off. Montane forests not only increase the total amount of precipitation but at the same time, they are crucial for stable surface water run-off. The total amount of precipitation increases not only by collecting horizontal precipitation but also by lower evaporation due to lower temperatures than in the lower parts of the catchment. Humus accumulations reduce run-off on a slope that subsequently does not cause erosion.

Submontane forests form the zonal vegetation between montane and floodplain ecosystems. They occur mostly on slopes with harmonious water balance. Soils are generally moderately permeable due to a balanced proportion of stoniness and fine-grained weathered particles. The formation of bogs is excluded on dominant, slanting slopes and more favorable temperatures that intensify soil respiration prevent excessive accumulation of surface humus. Higher clay content and lower humus accumulation distinguish water retention properties of submontane soils from montane soils. Submontane forests inhibit atmospheric precipitation only up to an amount corresponding with potential evapotranspiration, while continuous run-off along the surface as well as from the soil body occurs when WHC is exceeded.

Floodplain forests occur in a flat relief formed by floods. On the one hand, floods lay terraces; on the other hand, they tear down banks. The activity of rivers increases the diversity of soil properties, mostly at interfaces with zonal sites. The functionality of floodplain forests is determined by river water and waterlogging. The duration of the flood, the variability in the height of the river level and the fluctuation of the groundwater level induce differentiation of floodplain ecosystems. Extraordinary floods most significantly alter the dynamics of their development. The function of floodplain forests varies due to the lack of precipitation for evapotranspiration, which they are able to replace thanks to floods or high groundwater levels. The long-term decline of the soil water level at high evaporation can result in the replacement of the floodplain forest with the forest-steppe [65]. Floodplain forests with optimal soil moisture and high evaporation transpire almost 80% of potential evapotranspiration. This amount contains up to 70% of groundwater and 30% of precipitation. However, the transpiration of trees is not merely inhibited by the lack of soil water, but also by the lack of air during a prolonged flood [66].

2.2.4 Vulnerability of forest hydric functions

In Central Europe, the current health status of forest stands is closely linked to the climatic situation, particularly the availability of water for woody plants. Water in forest soils is a key part of the feedback relations, both in the soil–plant direction, currently mainly as a limiting eco-factor, and in the soil-landscape direction, in terms of the landscape water regime, water retention in the landscape and prevention of flood events.

Forest functions are threatened by dieback, fragmentation and transformation of tree species composition. The loss of forests leads to a decrease in evaporation, with cloud formation also declining. The decrease in cloud formation affects the whole catchment. Although the evaporation reduction should prevent soil moisture diminution, unlike evapotranspiration, it is not regulated by means of the vegetation cover, but merely by temperature alteration. A denuded land is easier to warm up, increasing biological activity and mineralization intensity. This occurs provided that removal of the stand component does not result in (frequent) waterlogging of a site, which would be limiting to the aerobic organisms at least until the lost functionality of the subsequent stand is restored. Soil without organic matter loses both water retention capacity and fertility. The decrease in forests is most distinctive in the lower parts of the catchment, which are more accessible, mostly non-waterlogged and more hydrologically suited for agriculture. Since the occurrence of precipitation also lowers in the spring-dependent parts of catchments as the cloud formation diminishes, the subsequent decline in river levels causes a decline in water supply to tributary-dependent parts of the catchment [67].

The greatest differences in the soil water-holding capacity are found between forested and treeless catchments. Flooding in forested catchment areas occurs in the aftermath of exceeding EDSP. Conversely, treeless catchments are affected by flash floods even after precipitation ˂30 mm. The protection of the water retention capacity of the catchment consists primarily in the prevalence of unbroken stretches of forests. Young open forest stands resemble treeless zones in terms of the water balance. Only closed stands over 20 years of age reach a water balance comparable to that of adults. Even though homogeneous forest stands provide hydric functionality similar to richly structured mixed forests, richly structured forests appear more resilient to climate change. Protecting the hydric functions of forests during climate change can be achieved in the following ways:

- Promotion of the transformation of tree species composition in favor of the natural state, with a natural proportion of trees within each stand type exceeding 50%

- Favoring understorey or small-scale differentiated farming to increase age and spatial diversity
- Maintaining a closed canopy to protect the soil surface, where understorey can be mined at the restoration stage without affecting the species diversity of vegetation
- Construction of a sufficiently dense transport network to minimize machinery driving through stands, giving priority to mining technologies that do not compact the upper soil horizons

Drought stress in forest stands has been shown to reduce both transpiration and the water content in plants [68, 69]. This occurs because of the loss of assimilation apparatus, thus reducing leaf area available for transpiration, but also because of the reduced availability of nutrients, which convert to a dehydrated state in a differentiated way [70]. At higher humidity, there is more Ca^{2+} and Mg^{2+} present in the soil solution, and at the same time, K^{+} is better released by mineralization processes. This is due to the size of the hydration envelope of the ions, which conditions their hydration energy for various nutrients in a differentiated way. This is necessary for the nutrient to be taken up by the plant. To hydrate diverse ions, different amounts of water molecules are needed, so potassium is absorbed at lower soil moisture than magnesium or calcium—two elements that frequently prove nutritionally deficient even though they may be at an optimal concentration in the soil.

2.3 Soil water and its relationship with groundwater

In the contemporary cultural landscape, the natural water cycle is, to a large extent, influenced (in other words, 'shortened') by vertical water movement within terrestrial systems. Consequently, communication within soil hydrological systems and the rock subsoil is impaired. The reasons will be explained in the following section.

The amount of water is distributed very unevenly in space and time on Earth. That is why there are problems with its lack in many regions. Redistribution of water in the landscape can be expressed by the fundamental elementary redistribution equation of water (this is also referred to as the balance equation [71]):

$$DP = IR + P - \Delta S - RO - ET \quad (1)$$

where DP: deep percolation; IR: irrigation; P: precipitation; RO: surface run-off; ET: evapotranspiration; ΔS: soil water storage.

On the basis of this balance equation, two basic hydrological cycles are identified: the large and small water cycles. In the water cycle (**Figure 3**), the main sources are precipitation and the surface, lateral and underground inflow in the hydrogeological collector. Water that falls on the soil surface immediately infiltrates the soil or, under conditions of insufficient infiltration capability and hydraulic conductivity, it drains or accumulates in micro-depressions of the relief (detention). Infiltrated water is redistributed in the soil body and remains below the soil surface, suspended in a capillary manner. Gravitational water then flows out of the area laterally (hypodermically) and migrates to the capillary fringe (see below), through which it percolates into an aquiferous hydrogeological collector. In relation to the vegetation, the water cycle is influenced by evapotranspiration and interception.

Soil, or more exactly the soil environment, is the main location of infiltration of water into the rocky underground environment. In general, this is the most important environment for the replenishment of groundwater supplies.

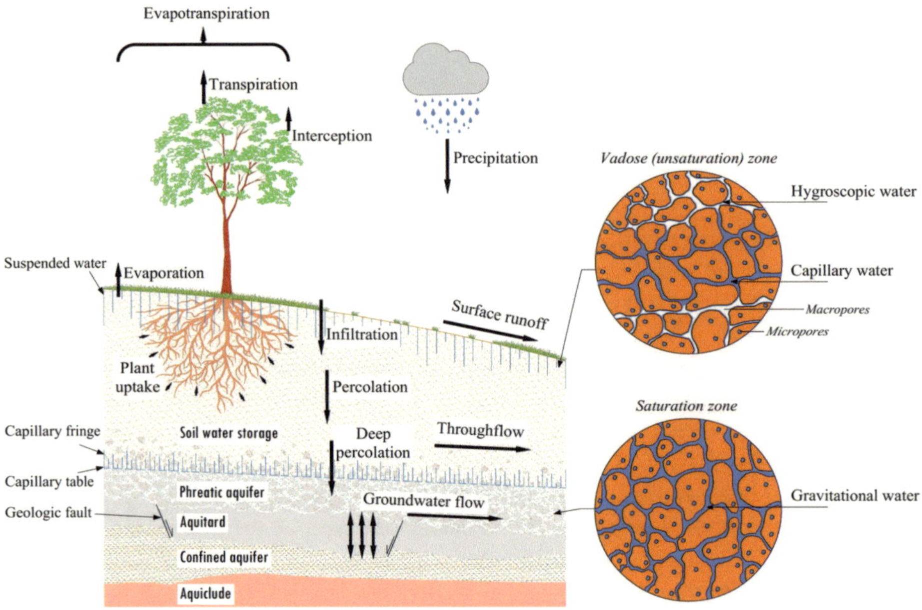

Figure 3.
The small water cycle in relation to geological subsoil: Communication of soil water and groundwater.

The subsurface water can be simply divided into soil water and groundwater. Although it is the same infiltrated surface water, these two divisions differ significantly from one another mainly in the ratio of forces acting on them. Soil water can be divided into three categories, namely adsorption, capillary and gravitational water.

The soil and rock environments can be classified into two zones in terms of saturation of the environment with water. The environment with the presence of air in pores may be termed the aerated unsaturated zone, where adsorption and pellicular water predominate, and gravitational water preponderates only for a limited period. On the contrary, the environment without the presence of air in pores (filled with water) is designated as a saturated aquiferous zone where gravitational water not bound by adsorption and capillary forces prevails. This water may be freely moving or maybe in the form of capillary water, filling small capillary pores.

The zone immediately adjacent to the aquiferous zone itself, that is, groundwater level, is the capillary fringe zone. Capillary water predominates in this zone. Adsorption water and, depending on the circumstances, gravitational water, may also be present. The capillary water completely fills capillary pores and is maintained by a capillary rise from the groundwater level in the zone. Capillary forces create a negative pore water pressure (under pressure). Thus, water cannot be collected from the environment and responds merely to groundwater level fluctuations. From hydrogeology and groundwater hydraulics, the capillary fringe zone can be included in the unsaturated (non-aquiferous) zone. Contrarily, in hydropedology, we work with the capillary fringe zone as with the saturated zone, which significantly affects the physico-chemical properties of the soil and is important in terms of the water supply of the soil environment in agriculture.

In terms of replenishing groundwater reserves by infiltration, gravitational water is the most significant. Gravitational water is used during infiltration, especially for the area of the rock environment above the groundwater level, that is the unsaturated environment. This includes the area between the groundwater level and the subsurface soil-water zone. The capillary fringe zone can also be ranked in this category.

The principle of water infiltration into the rock environment in the unsaturated zone can be expressed by gravitational and water potentials. In particular, infiltration depends on the characteristics of the particular soil or weathered particles (grain size, structure, organic matter content, geological activity, stratigraphy, etc.). Infiltration is determined experimentally for each specific soil type. For this purpose, moisture curves are used, which express the relationship between capillary pressure and moisture. The curves differ (hysteresis of the curves) when the soil is filled with water and when it dries.

Vertical flow of infiltrated water through the soil medium is such that during infiltration, pores in the upper soil layer become increasingly saturated with rainwater until saturation of the water-holding capacity is reached, whereupon the saturated zone shifts gravitationally deeper in the soil profile. This occurs because semi-capillary and non-capillary pores are systematically filled with water above the hydrolimit of the water-holding capacity and water moves with gravity in terms of saturated flow according to Darcy's law. As rainwater supply ends, due to termination of the particular rain event, saturation is reduced, and gravitational flow of water slows down and gradually begins to be controlled by the hydraulic conductivity of the particular type of the soil. Water dissipated in the environment and movement is practically stopped. If the rainwater supply is sufficient, infiltrated water may eventually reach the groundwater level, which is progressively raised. Due to gravitational drainage into the body of groundwater, the saturation of the soil environment gradually decreases, and the unsaturated zone is created again.

The process of infiltration through the soil environment substantially affects the quality of the infiltrated water, both positively, when it can significantly reduce pollution and thus protect groundwater against chemical or microbial contamination, but also negatively, in the case of contaminated soils (by means of anthropogenic activity, such as the. Enormous doses of industrial fertilizers applied to agricultural soils). Here, the contaminated infiltrated water can lead to the deterioration of the groundwater reservoir.

At present (i.e. in this current episode of anthropogenically driven climate change), it is of utmost importance to maintain the soil environment in as favorable as possible a condition in terms of enabling infiltration of rainwater into the soil environment or, more precisely, into the groundwater collector. The principal negative factors include soil compaction, the loss of soil structure and the reduction of organic matter content in the soil. These three factors significantly reduce the water-holding capacity of the soil, that is, the ability to retain and gradually release water, either in the form of evapotranspiration or infiltration into the groundwater reservoir. Vast impermeable anthropogenic surfaces (asphalt, concrete, roofs, etc.) also inflict a significant reduction in infiltration.

Nowadays, it is highly desirable to ensure infiltration of rainwater from these areas by appropriate technical and biotechnical measures, thus preventing their rapid surface or sub-surface run-off. Groundwater recharge in the Central European region historically took place in the colder half of the year, mainly from snowmelt. In this region this represented 3–4 months a year, when the zone between soil water and groundwater level was saturated and thus the regional groundwater reserves were continually replenished.

In the last 20 years, probably due to climate change, but also relevant alterations in landscape utilization, the saturation period of this zone has been significantly shortened and, consequently, there has been limited replenishment of groundwater supplies. A key role is played by noticeably lower snow reserves in the winter months, the overall temperature elevation during the year (i.e. increased evapotranspiration), and changes in rainfall distribution (accumulation of rainfall and decreased soil absorption capacity). Groundwater recharge is thus usually carried out during longer term, higher rainfall events. In the case of torrential rain, the surface zone is rapidly saturated and hence minimum infiltration, and erosive strong surface run-off occurs. Contrarily, during long-term moderate rain, the entire transitional zone gradually saturates to the groundwater level, and thus its reserves are replenished.

Groundwater reservoirs are also replenished at tectonic faults (fractures). The entire soil body need not be saturated within the process of infiltration, but gravitational water can flow because of fissure permeability, replenishing the groundwater reserves.

2.4 The soil-forest-water-civilization nexus

2.4.1 The elements of life

The soil-forest-water-civilization nexus has never been more important than at present. The Ancient Greeks recognized four basic elements of life: fire, water, air and soil. Yet throughout history, perturbation of the hydrosphere, atmosphere and geosphere has created huge issues for humanity and the rest of the Biosphere.

Trees are an essential component of most ecosystems on our planet, and the forests of the world play key roles in the hydrological cycle, nutrient cycles and the carbon cycle. Deforestation undermines ecosystem function upon which we rely for our very survival. Forests are major contributors to rainfall, with the Amazon rainforest producing some 70% of precipitation in the Rio de Plata river basin [72]. Forests also play a crucial role in temperature regulation, not only as repositories for carbon, but in terms of evapotranspiration and the production of microbial flora and biogenic volatile organic compounds which act as condensation nuclei for cloud formation and rain events. It is estimated that deforestation may account for as much as 18% of current global warming [73]. Forests purify surface and ground water [74]. Deforestation also reduces soil structure and organic carbon content, negatively impacting the water-holding capacity [75]. Environmental degradation leads to economic collapse and social instability [76]. Healthy forests and healthy soils are inextricably linked. Deforestation has three significant impacts: soil erosion, soil salinization and eutrophication.

2.4.2 Soil erosion

The incredible diversity of the biosphere in its many forms speaks to a complex foundation upon which such a magnificent edifice is built. Yet terrestrial ecosystems are almost entirely dependent upon a thin, living skin, stretching across some fifty million square kilometers, but with a mean depth of only 15 cm: the soil. Most plants need soil, and plants form the basis of most terrestrial food chains. Yet in the last 150 years, we have lost 50% of the planet's topsoil through soil erosion. Lester W. Brown, the President of the Earth Policy Institute, has written that civilization can survive the loss of its oil reserves, but it cannot survive the loss of its soil reserves [77].

Soil erosion is not a new problem. Plato bemoaned the fact that the soil of Greece was, by his own time, eroding, observing that 'what now remains compared with what then existed is like the skeleton of a sick man, all the fat and soft earth having wasted away, and only the bare framework of the land being left' (in Glacken [78]). Around 60 BC, Lucretius, the philosopher and poet, recognized the seriousness of soil exhaustion in Italy. He thought that the Earth itself was dying [79]. A comprehensive review of the historical significance of soil erosion and the contribution of deforestation to this can be found in Dotterweich [80].

Accelerated erosion has been occurring in Britain since the first clearances of primeval forest 5000–6000 years ago [81]. While early human agriculturalists used hand-held tools, maintaining a rough surface, allowing infiltration, later iron tools smoothed the surface, leading to run-off and erosion. By medieval times, many European villages had been abandoned as a result of soil erosion, elevating food prices due to crop failure and leading to social instability [82]. Today, 751 million ha of the planet's soil has been severely eroded [83]. Overgrazing by livestock and intensive agricultural practice has led to huge swathes of erosion. But deforestation has been one of the most significant contributors to the erosion crisis facing the planet. One and a half million square kilometers of dense tree cover were lost between 2000 and 2012 [84].

Shallow tree roots bind soil aggregates, increasing soil cohesion, while protecting against surface wash erosion. Deeper roots anchor the regolith to the bedrock, preventing landslides, debris flows and mudflows. Trees also reduce the load from lower soil moisture through evapotranspiration [85].

Soil production takes many years, and today losses far exceed formation. In China, the soil is being lost 54 times faster than it is being formed, leading to huge economic and social insecurity. In the case of China, soil loss accounts for the loss of 42 billion dollars per year, impacting 170 million people [86].

It is thought that the Babylonian and Sumerian kingdoms collapsed due to soil erosion, blocking irrigation systems [87]. Once the soil is gone, the risk of flooding after heavy rain increases dramatically. The trees form a crucial link in the hydrological cycle, shifting water from the soil back to the atmosphere.

Wind erosion is an equally serious threat to humanity. The Dust Bowl in the USA stands as a striking example, where a 10-year collapse in agriculture was due to soil erosion driven by agricultural mismanagement in the 1930s. On Black Sunday, 14 April 1935, the sunlight was blocked out by the dust, when three million tonnes of topsoil from the Midwest was blown into the atmosphere. The Dust Bowl forced around two and a half million people to flee from their mid-west farms and head to California.

2.4.3 Eutrophication

Soil erosion contributes to another major threat to our planet, eutrophication. Eutrophication is caused by nutrients being washed into the hydrosphere from the soil. Soil itself is a nutrient bomb, and so erosion delivers huge amounts of nutrients into streams, rivers, lakes and the oceans, leading to hypoxia, cyanobacterial blooms, toxic red tides and fish death. In Europe, Asia and North America, 50% of freshwater bodies are now eutrophic, while dead zones are a regular occurrence in the oceans, devastating fish populations. In the US alone, eutrophication costs around 2 billion dollars each year [88].

2.4.4 Soil salinization

Deforestation also leads to soil salinization. Currently, 25% of the world's cropland is affected, while in Africa, this figure is 50% [89]. By 2050 it is estimated that some 50% of cropland will have productivity halved due to build-up of salt in the surface soil [90]. Nagendran [91] observes that salinization is the most striking effect of agriculture in all parts of the world. Soil salinization is very difficult to reverse.

Salinization is a particular threat to Australian agriculture, given that most of the country is desert. In the Murray Darling Basin, 63% of the forested area has been converted to cropland in the last 200 years [92]. This has led to increased downward water fluxes below the root zone by one to two orders of magnitude [93] because the trees are no longer performing their role as water shifters from soil to atmosphere. This has resulted in a rapid rise in the groundwater table at a rate of~1 m year^{-1}, leading to the salinization of some 5.7 million ha of farmland, devastating harvests [94].

Similar large-scale salinization events have been recorded in California, north-west India and much further back in time, in Ancient Mesopotamia [95–97].

2.4.5 The biotic pump

Finally, deforestation leads to huge changes in the rainfall distribution patterns on our planet. The biotic pump theory [98] proposes that evapotranspiration creates lower pressure above forest canopies, drawing in moist air from the oceans, and supplying precipitation far inland. The reduction in evapotranspiration as a result of deforestation leads to an increase in the height of the convective boundary layer because of the stronger sensible heat flux over pastures. This is less conducive to rainfall formation. Deforestation is thought to have contributed 60% to the drought conditions that led to the collapse of the Mayan empire [99].

Much like climate destabilization, the biotic pump acts across national boundaries, requiring international collaboration. If inland nations carry out significant deforestation, the impacts are not only felt within that nation, in terms disruption to the local hydrological cycle, exacerbating flood risks, landslides, soil erosion and water purity, but also in nations that lie between the oceans and the deforested region, as the pressure gradient is no longer as strong, reducing the strength of the pump.

Critics of the biotic pump theory have argued that air movements as a result of condensation are multi-directional, representing an isotropic (uniform in all directions) process and this means that there will not be any uni-directional, net flow from ocean to continental landscapes [100]. In this orthodox approach, mass air movements alone drive the hydrological cycle across latitudinal cells set up by temperature gradients due to the uneven heating from the sun as a result of the axial tilt and curvature of the Earth.

However, it has been demonstrated experimentally that condensation can trigger anisotropic, uni-directional flow, supporting the biological pump theory [101, 102]. Sheil [103] points out that disruption of the biological pump through deforestation can lead to dramatic, non-linear transitions in local climate, from wet to dry regimes. Interestingly, reforestation can lead to a similarly dramatic transition in the opposite direction, from a dry to a wet local climate regime [103]. However, there is no guarantee that reforestation will return the region to an identical ecological state as that prior to deforestation, as species may have suffered extinction, and recolonization routes may no longer exist.

2.4.6 Regime shifts

Of greater concern yet is the fact that such widescale changes resulting from deforestation and the destabilization of the soil-water relationship may lead to regime shifts. Lees et al. [104] define regime changes as abrupt changes on several trophic levels, leading to rapid ecosystem reconfiguration between alternative states. Both structures and processes are transformed and such changes, in turn, result in significant alterations in ecosystem services [105, 106]. Complex non-linear systems, such as ecosystems, become vulnerable to phase shifts, where relatively small changes in an already stressed system can result in the irreversible collapse of the system, switching, for example, from a wet forested state to a dry savanna, and creating an alternative equilibrium, with devastating consequences [107–109]. Such shifts are more likely to occur as anthropogenic perturbation increases [110].

Of additional concern is the reality that ecosystems are interconnected to other ecosystems, to such an extent that a regime change in one part of the biosphere can catalyze changes in other ecosystems. One example relates to regime changes in the Arctic, wherein sea-ice changes lead to reorganization of tropical convection that in turn triggers an anticyclonic response over the North Pacific, resulting in significant drying over California [111], potentially leading to regime change. Ecosystems are sub-systems, not isolated systems. Thus, changes run throughout the biosphere, impacting on all levels of organization, in non-linear ways. We would expect this in any self-organizing system, where feedback dictates context and change. One such conduit is the soundscape, wherein ecological simplification can lead to radical transitions at the ecosystem level facilitated by the absence of audible cues [112]. Another feedback conduit is the hydrological cycle, and forests play a central role here. Interfering with water relations can have huge impacts on regime stability and the spread of regime shifts across the biosphere [113].

3. Conclusion

Forest soil water balance plays an essential, central role in ecosystem functionality. The modification of water balance within forests can enhance self-regulation of all ecosystems in a landscape, but intensive, anthropogenic landscape transformation can negatively impact it. Human activities, such as deforestation, have had damaging impacts on evaporation, precipitation and run-off. The protection of forest water balance has been highlighted as a priority through coordinated research based on analysis of soil properties and ecosystem function restoration. Underpinning any hope of achieving this lies the urgency of attaining a sustainable relationship between human needs and natural resources.

Thus, we see that forests are essential components in both the hydrological cycle and in soil functionality, while also playing a crucial role in the carbon cycle. Forests, much like soil and water, are currently under-appreciated by the human race, yet our futures rely on their restoration and respect. Kravčík [114] have called for a new paradigm in order to rescue humanity from a crisis beyond our imagining: regime shifts and the functional collapse of the terrestrial and aquatic ecosystems. Such a paradigm no longer views water as an isolated entity, a fixed renewable resource and having little to do with the suite of environmental crises facing us, along with the coming economic and societal collapse undoubtedly awaiting us on our current trajectory. Instead, they call for a prioritization of the restoration of the water balance at all levels, but particularly at the level of the small water cycle. Intrinsic to this is healthy soils and healthy forests. The soil-forest-water-civilization nexus must urgently be understood as a synergy, connected and united within the Earth system if we are to find a constructive way ahead and a place for our own sub-species within the biosphere.

Acknowledgements

This chapter was supported with the institutional support of Mendel University in Brno financed from the institutional support of the development of the research organization provided by the Ministry of Education, Youth and Sports, Czech Republic.

Author details

Aleš Kučera[1], Pavel Samec[1], Aleš Bajer[1], Keith Ronald Skene[2], Tomáš Vichta[1], Valerie Vranová[1]*, Ram Swaroop Meena[3] and Rahul Datta[1]

1 Department of Geology and Pedology, Faculty of Forestry and Wood Technology, Mendel University in Brno, Brno, Czech Republic

2 Biosphere Research Institute, Letham, Angus, United Kingdom

3 Department of Agronomy, Institute of Agricultural Sciences, Banaras Hindu University, Varanasi, India

*Address all correspondence to: vranova@mendelu.cz

References

[1] Nieminen TM, Derome K, Meesenburg H, De Vos B. Soil solution: Sampling and chemical analyses. In: Ferretti M, Fischer R, editors. Forest Monitoring Methods for Terrestrial Investigations in Europe with an Overview of North America and Asia. Developments in Environmental Science. Vol. 12. Oxford, UK: Elsevier; 2013. pp. 301-315

[2] Bruijnzeel LA, Mulligan M, Scatena FN. Hydrometeorology of tropical montane cloud forests: Emerging patterns. Hydrological Processes. 2011;**25**:465-498. DOI: 10.1002/hyp.7974

[3] Tognetti R. Trees harvesting the clouds: Fog nets threatened by climate change. Tree Physiology. 2015;**35**: 921-924. DOI: 10.1093/treephys/tpv086

[4] Lindroos AJ, Derome J, Mustajärvi K, Nöjd P, Beuker E, Helmisaari HS. Fluxes of dissolved organic carbon in stand throughfall and peercolation water in 12 coniferous stands on mineral soils in Finland. Boreal Environment Research. 2008;**13**:22-34

[5] Armbruster MS, Seegert J, Feger K. Effects of changes in tree species compositioin on water flow dynamics—Model application and their limitations. Plant and Soil. 2004;**264**:13-24

[6] Cermak J, Nadezhdina N. Chapter 4: Field studies of whole-tree leaf and root distribution and water relations in several European forests. In: Bredemeier et al., editors. Forest Management and the Water Cycle: An Ecosystem-Based Approach, Ecological Studies, Part 1. Vol. 212. Dordrecht, Germany: Springer; 2011. pp. 65-88

[7] Kutílek M. Soil Science for Water Management. Prague, In Czech: SNTL-ALFA; 1978

[8] Pirastru M, Castellini M, Giadrossich F, Niedda M. Comparing the hydraulic properties of forested and grassed soils on an experimental hillslope in a Meditarranean environment. Procedia Environmental Sciences. 2013;**19**:341-350

[9] Hao M, Zhang J, Meng M, Chen HYH, Guo X. Imipacts of changes in vegetation on saturated hydraulic conductivity on soil in subtropical forests. Scientific Reports. 2019;**9**:8372

[10] Zanella A, Jabiol B, Ponge JF, Sartorid G, De Waale R, Van Delfte B, et al. European Humus Forms Reference Base. HAL: HAL-00541496. 2011. p. 56

[11] Birch LG. The effect of soil drying on humus decomposition and nitrogen availability. Plant and Soil. 1958;**10**:9-31

[12] Minderman G. Mull and mor (Mülleer-Hesselman) in relation to the soil water regime of a forest. Plant and Soil. 1960;**13**(1):1-27

[13] Sazonova T, Pridacha V, Olchev A. The water regime of silver (*Betula pendula* Roth) and Karelian (*Betula pendula* var. carelica) birches under sufficient and limited soil moisture conditions. Geophysical Research Abstracts. 2012;**14**:7168

[14] Lobet G, Couvreur V, Meunier F, Javaux M, Draye X, et al. Plant Physiology. 2014;**164**:1619-1627

[15] Porporato A, Daly E, Itube IR. Soil water balance and ecosystem response to climate change. The American Naturalist. 2004;**164**(5):625-632

[16] Matejka F, Rožnovský J, Hurtalová T, Janouš D. Effect of soil drought on evapotranspiration of a young spruce forest. Journal of Forest Science. 2002;**48**(4):166-172

[17] Cienciala E, Kucera J, Ryan MG, Lindroth A. Water flux in boreal forest during two hydrologically contrasting

years; species specific regulation of canopy conductance and transpiration. Annals of Forest Science. 1998;**55**:47-61

[18] Farkas C, Gelybó G, Bakasci Z, Horel Á, Hagyó A, Dobor L, et al. Impact of expected climate change on soil water regime under diferent vegetation conditions. Biologia. 2014;**69**(11):1510-1519

[19] Brooks N, Adger N. Assessing and enhancing adaptive capacity. In: Lim B, editor. Adaptation Policy Frameworks for Climate Change: Developing Strategies, Policies and Measures. Cambridge: UNDP and Cambridge University Press; 2004. pp. 165-181

[20] Dambrine I, Carisey N, Pollier B, Granier A. Effect of drought on the yellowing status and the dynamics of mineral elements in the xylem sap of declining spruce (Pucea ables L.). Plant and Soil. 1993;**150**(2):303-306. DOI: 10.1007/BF00013028

[21] Tóth B, Makó A, Guadagnini A, Tóth G. Water retention of salt-affected soils: Quantitative estimation using soil survey information. Arid Land Research and Management. 2012;**26**(2):103-121

[22] Ghestem M, Sidle R. The influence of plant root systems on subsurface flow: Implications for slope stability. Bioscience. 2011;**61**:869-879

[23] Ilek A, Kucza J, Szostek M. The effect of stand species composition on wateer storage capacity of the organic layers of forest soils. European Journal of Forest Research. 2015;**134**:187-197

[24] Brady NC, Weil RR. The Nature and Properties of Soils. 13th ed. New Jersey: Prentice Hall; 2002. p. 960. ISBN: 0-13-016763-0

[25] White RE. Principles and Practice of Soil Science, the Soil as a Natural Resource. 4th ed. UK: Bleckwell Publishing; 2006. p. 363. ISBN-13: 978-0-632-06455-2

[26] Vavříček D, Kučera A. Základy lesnického půdoznalství a výživy lesních dřevin (in Czech). Kostelec nad Černými Lesy: Lesnická práce; 2017. p. 364. ISBN: 978-80-7458-103-8

[27] Lozano-Baez SE, Cooper M, de B Ferraz SF, Rodrigues RR, Lassabatere L, Castellini M, et al. Assessing water infiltration and soil water repellency in Brazilian Atlantic forest soil. Applied Sciences. 2020;**10**:1950

[28] Wang W, Zhang H, Li M, Cheng J, Wang B, Lu W. Infiltration characteristics of water in forest soils in the simian mountains, Shongquing City, southwestern China. Frontiers of Forestry in China. 2009;**4**:338-343

[29] Harden CP, Scruggs PD. Infiltration on mountain slopes: A comparison of three environments. Geomorphology. 2003;**55**:5-24

[30] USDA-NRCS. Soil Taxonomy, A Basic System of Sol Classification for Making and Interpreting Soil Surveys. 2nd ed. Washington, DC; 1999. p. 869

[31] Dahiya IS, Dahiya DJ, Kuhad MS, Karwasra SPS. Statistical equations for estimating field capacity, wilting point and availabe water capacity of soils from their saturation percentage. The Journal of Agricultural Science. 1988;**110**(3):515-520

[32] Mbah CN. Determination the field capacity, wilting point and available water capacity of come southeast nigerian soils using soil saturation from capillary rise. Nigerian Journal of Biotechnology. 2012;**24**:41-47

[33] Silva BM, da Silva ÉA, de Oliveira GC, Ferreira MM, Serafim ME. Plant-available soil water capacity: Estimation methods and implications. Revista Brasileira de Ciência do Solo. 2014;**38**:464-475

[34] Teepe R, Dilling H, Beese F. Estimating water retention curves of

forest soils from soil texture and bulk density. Journal of Plant Nutrition and Soil Science. 2003;**166**:111-119

[35] Cary JW, Hayden CW. An index for soil pore size distribution. Geoderma. 1973;**9**(4):249-256

[36] Smucher AJM, Park EJ, Dorner J, Horn R. Soil micropore development and contributions to soluble carbon transport within macroaggregates. Vadose Zone Jurnal. 2007;**6**:282-290

[37] Zhang F, Cui YJ, Ye WM. Distinguishing macro- and micro-pores for materials with different pore populations. Géotechnique Letters. 2018;**8**:1-9

[38] Lu N, Likos WJ. Rate of capillary rise in soil. Journal of Geotechnical and Geoenvironmental Engineering. 2004;**130**(6):646-650

[39] Perevill KI, Sparrow LA, Reuter DJ, editors. Soil Analysis: An Interpretation Manual. Collingwood, Australia: CSIRO Publishing; 1999. ISBN: 0-643-06376-5

[40] Pielke RA. Land use and climate change. Science. 2005;**310**:1625-1626. DOI: 10.1126/science.1120529

[41] Muys B, Nyssen J, du Toit B, Vidale E, Prokofieva I, Mavsar R, Palahi M. Water-related ecosystem services of forests: Learning from regional cases. In: Katila P, Galloway G, de Jong W, Pacheco P, Mery G. editors. Forests under Pressure—Local Responses to Global Issues. IUFRO World Series 32; 2014. p. 423-440

[42] Haines-Young R, Potschin M. The links between biodiversity, ecosystem services and human well-being. In: Raffaelli D, Frid C, editors. Ecosystem Ecology: A New Synthesis. Cambridge: Cambridge University Press; 2010. pp. 110-139

[43] Dufrêne E, Davi H, Francois C, Le Maire G, Le Dantec V, Granier A. Modelling carbon and water cycles in a beech forest part I: Model description and uncertainty analysis on modelled NEE. Ecological Modelling. 2005;**185**:407-436. DOI: 10.1016/j.ecolmodel.2005.01.004

[44] Muys B, Ceci P, Hofer T, Veith C. Towards integrated ecological, socio-economic and hydrological management. In: Birot Y, Gracia C, Palahí M, editors. Water for Forests and People in Mediterranean Region—A Challenging Balance. Avignon: EFI; 2011. pp. 105-113

[45] Trenberth KE. Changes in precipitation with climate change. Climate Research. 2011;**47**:123-138. DOI: 10.3354/cr00953

[46] Ellison D. Forests and Water. New York: Global Forest Goals, United Nation Forum on Forests; 2018. Available from https://www.un.org/esa/forests/wp-content/uploads/2018/04/UNFF13_BkgdStudy_ForestsWater.pdf [Accessed: 27 February 2020]

[47] Lal R. Forest soils and carbon sequestration. Forest Ecology and Management. 2005;**220**:242-258. DOI: 10.1016/j.foreco.2005.08.015

[48] Rejšek K. Seen among the crowd: The organism and soil ecosystem. Phytopedon (Bratislava). 2004;**3**:18-21

[49] Paoletti E, Schaub M, Matyssek R, Wisser G, Augustaitis A, Bastrup-Birk AM, et al. Advances of air pollution science: From forest decline to multiple-stress effects on forest ecosystem services. Environmental Pollution. 2010;**158**:1986-1989. DOI: 10.1016/j.envpol.2009.11.023

[50] Peterman W, Bachelet D. Climate change and forest dynamics: A soils perspective. In: Hester RE, Harrison RM, editors. Soils and Food Security. Issues in Environmental Science and Technology. Vol. 35. 2012. pp. 158-182. DOI: 10.1039/1465-1874

[51] Gauger T, Anshelm F, Schuster H, Erisman JW, Vermeulen AT, Draaijers GPJ, Bleeker A, Nagel HD. Mapping of Ecosystem Specific Long-Term Trends in Deposition Loads and Concentrations of Air Pollutants in Germany and their Comparison with Critical Loads and Critical Levels. Umweltbundesamt, Berlin (Final Report). 2002. Available from: https://www.nav.uni-stuttgart.de/img/critical_loads/EB_29942210_T1.pdf [Accessed: 27 February 2020]

[52] Piedallu C, Gégout J-C, Bruand A, Seynave I. Mapping soil water holding capacity over large areas to predict potential production of forest stands. Geoderma. 2011;**160**(3-4):355-366. DOI: 10.1016/j.geoderma.2010.10.004

[53] Cools N, De Vos B. Part X: Sampling and analysis of soil. In: UNECE ICP Forests Programme Coordinating Centre, editor. Manual on Methods and Criteria for Harmonized Sampling, Assessment, Monitoring and Analysis of the Effects of Air Pollution on Forests. Eberswalde: Thünen Institute of Forest Ecosystems; 2016. pp. 1-29. Available from: https://www.icp-forests.org/pdf/manual/2016/ICP_Manual_2016_01_part10.pdf [Accessed: 27 February 2020]

[54] Goldberg S, Lebron I, Suarez DL. Soil colloidial behavior. In: Sumner ME, editor. Handbook of Soil Science. Boca Raton – Abingdon: CRC Press; 2000. pp. 195-240

[55] Yan C, Zhao WL, Wang Y, Yang Q, Zhang Q, Qiu G. Effects of forest evapotranspiration on soil water budget and energy flux partitioning in a subalpine valley of China. Agricultural and Forest Meteorology. 2017;**246**:207-217. DOI: 10.1016/j.agrformet.2017.07.002

[56] Dangles O, Malmqvist B, Laudon H. Naturally acid freshwater ecosystems are diverse and functional: Evidence from boreal streams. Oikos. 2004;**104**:149-155. DOI: 10.1111/j.0030-1299.2004.12360.x

[57] Booth EG, Loheide SP II. Hydroecological model predictions indicate wetter and more diverse soil water regimes and vegetation types following floodplain restoration. Journal of Geophysical Research. 2012;**117**:G02011. DOI: 10.1029/2011JG001831

[58] Olson DM, Dinerstein E, Wikramanayake ED, Burgess ND, Powell GVN, Underwood EC, et al. Terrestrial ecoregions of the world: A new map of life on earth. Bioscience. 2001;**51**:933. DOI: 10.1641/0006-3568(2001)051[0933,TEOTWA]2.0.CO;2

[59] Fischer G, Nachtergaele F, Prieler S, van Velthuizen HT, Verelst L, Wiberg D. Global Agro-Ecological Zones Assessment for Agriculture (GAEZ 2008). Rome: IIASA, FAO; 2008

[60] Samec P, Vavříček D, Kučera A. Povodně a hydrické potenciály lesních půd v Moravskoslezském kraji (in Czech). In: Samec P, editor. Změny klimatu a lesnictví, ČZU v Praze. 2008. pp. 91-124

[61] Trenberth KE, Smith L, Qian T, Dai A, Fasullo J. Estimates of the global water budget and its annual cycle using observational and model data. Journal of Hydrometeorology. 2007;**8**:758-769. DOI: 10.1175/JHM600.1

[62] Šach F, Kantor P, Černohous V. Stanovení evapotranspirace mladého smrkového a bukového porostu metodou kontinuálního měření objemové vlhkosti v půdním profilu (in Czech). In: Jurásek A, Novák J, Slodičák M, editors. Stabilizace Funkce lesa v Biotopech Narušených Antropogenní Činností v Měnících se Podmínkách Prostředí. Opočno: VÚLHM; 2000. pp. 525-536

[63] Švihla V, Černohous V, Kulhavý Z, Šach F. Retence srážkové vody lesní půdou v horském povodí (in Czech). In: Neuhöferová P, editor. Meliorace

v lesním hospodářství a v krajinném inženýrství. Praha—Kostelec nad Černými lesy: ČZU v Praze, VÚMOP. 2006. pp. 35-44

[64] Kantor P, Šach F. Hydrická účinnost mladých náhradních porostů smrku omoriky a břízy bradavičnaté (in Czech). Lesnictví. 1988;**34**:1017-1040

[65] Škvarenina J, Tomlain J, Križová E. Klimatická vodní bilance vegetačních stupňů na Slovensku (in Slovak). Meteorologické Zprávy. 2002;**55**:103-109

[66] Klimo E, Hager H, Matić S, Anić I, Kulhavý J, editors. Floodplain Forests of the Temperate Zone of Europe. Lesnická práce: Kostelec nad Černými lesy; 2008. p. 623

[67] Lawton RO, Nair US, Pielke R Sr, Welch RM. Climatic impact of tropical lowland deforestation on nearby montane cloud forests. Science. 2001;**294**:584-587. DOI: 10.1126/science.1062459

[68] Saxe H, Cannell MGR, Johnsen Y, Ryan MG, Vourlitis G. Tree and forest functioning in response to global warming. The New Phytologist. 2001;**149**(3):369-400. DOI: 10.1046/j.1469-8137.2001.00057.x

[69] Solberg S. Summer dought: A driver for crown condition and mortality of Norway spruce in Norway. Forest Pathology. 2004;**34**(2):93-104. DOI: 10.1111/j.1439-0329.2004.00351.x

[70] Smit B, Wandel J. Adaptation, adaptive capacity and vulnerability. Global Environmental Change. 2014;**16**:282-292. DOI: 10.1016/j.gloenvcha.2006.03.008

[71] Shukla SP. Water Management and Hydraulic Engineering in India, C. 600 BCE-CE 1200. Pentagon Press; 2014. p. 162. ISBN: 8182747414, 9788182747418

[72] Van der Ent RJ, Savenije HH, Schaefli B, Steele-Dunne SC. Origin and fate of atmospheric moisture over continents. Water Resources Research. 2010;**46**(9):1-12. W09525. DOI: 10.1029/2010WR009127

[73] Alkama R, Cescatti A. Biophysical climate impacts of recent changes in global forest cover. Science. 2016;**351**(6273):600-604. DOI: 10.1126/science.aac8083

[74] Neary DG, Ice GG, Jackson CR. Linkages between forest soils and water quality and quantity. Forest Ecology and Management. 2009;**258**(10):2269-2281. DOI: 10.1016/j.foreco.2009.05.027

[75] Lal R. Deforestation and land-use effects on soil degradation and rehabilitation in western Nigeria. I. Soil physical and hydrological properties. Land Degradation and Development. 1996;**7**(1):19-45. DOI: 10.1002/(SICI)1099-145X(199603)7:1<19::AID-LDR212>3.0.CO;2-M

[76] Skene K, Murray A. Sustainable Economics: Context, Challenges and Opportunities for the 21st-Century Practitioner. 1st ed. Oxford: Routledge; 2017. p. 462. DOI: 10.4324/9781351286206

[77] Brown LR. The New Geopolitics of Food. Food and Democracy [Internet]2011. p. 23. Available from: http://www.foreignpolicy.com/articles/2011/04/25/the_new_geopolitics_of_food [Accessed: 19 February 2020]

[78] Glacken CJ. Traces on the Rhodian Shore: Nature and Culture in Western Thought from Ancient Times to the End of the Eighteenth Century. 1st ed. Berkeley, CA: University of California Press; 1976. p. 763

[79] Green WM. The dying world of Lucretius. American Journal of Philology. 1942;**63**:51-60

[80] Dotterweich M. The history of human-induced soil erosion: Geomorphic legacies, early descriptions

and research, and the development of soil conservation - a global synopsis. Geomorphology. 2013;**201**:1-34. DOI: 10.1016/j.geomorph.2013.07.021

[81] Boardman J. Soil erosion in Britain: Updating the record. Agriculture. 2013;**3**:418-442. DOI: 10.3390/agriculture3030418

[82] Fraser EDG. Can economic, land use and climatic stresses lead to famine, disease, warfare and death? Using Europe's calamitous 14th century as a parable for the modern age. Ecological Economics. 2011;**70**:1269-1279. DOI: 10.1016/j.ecolecon.2010.02.010

[83] Lal R. Soil erosion and the global carbon budget. Environment International. 2003;**29**:437-450. DOI: 10.1016/S0160-4120(02)00192-7

[84] Hansen MC, Potapov PV, Moore R, Hancher M, Turubanova SA, Tyukavina A, et al. High-resolution global maps of 21st-century forest cover change. Science. 2013;**342**:850-853. DOI: 10.1126/science.1244693

[85] Reubens B, Poesen J, Danjon F, Geudens G, Muys B. The role of fine and coarse roots in shallow slope stability and soil erosion control with a focus on root system architecture: A review. Trees. 2007;**21**:385-402. DOI: 10.1007/s00468-007-0132-4

[86] Skene KR. Artificial Intelligence and the Environmental Crisis. 1st ed. New York: Routledge; 2019. p. 276. DOI: 10.1201/9780429055676

[87] Hillel DJ. Out of the Earth-Civilization and the Life of the Soil. 1st ed. New York, NY: The Free Press; 1991. p. 310

[88] Dodds WK, Bouska WW, Eitzmann JL, Pilger TJ, Pitts KL, Riley AJ, et al. Eutrophication of US freshwaters: Analysis of potential economic damages. Environmental Science & Technology. 2009;**43**:12-19. DOI: 10.1021/es801217q

[89] Ceuppens J, Wopereis MCS. Impact of non-drained irrigated rice cropping on soil salinization in the Senegal River Delta. Geoderma. 1999;**92**:125-140. DOI: 10.2136/sssaj1997.03615995006100040019x

[90] Bartels D, Sunkar R. Drought and salt tolerance in plants. Critical Reviews in Plant Sciences. 2005;**24**:23-58. DOI: 10.1080/07352680590910410

[91] Nagendran R. Agricultural waste and pollution. In: Letcher TM, Valleropp DA, editors. Waste: A Handbook for Management. 1st ed. Amsterdam: Elsevier; 2011. pp. 341-355. DOI: 10.1016/B978-0-12-381475-3.10034-8

[92] Walker J, Bullen F, Williams BG. Ecohydrological changes in the Murray-Darling basin. I. the number of trees cleared over two centuries. Journal of Applied Ecology. 1993;**30**:265-273. DOI: 10.2307/2404628

[93] Gordon L, Dunlop M, Foran B. Land cover change and water vapour flows: Learning from Australia. Philosophical Transactions of the Royal Society of London. Series B, Biological Sciences. 2003;**358**:1973-1984. DOI: 10.1098/rstb.2003.1381

[94] National Land and Water Resource Audit. Australian Dryland salinity assessment 2000: Extent, impacts, processes, monitoring and management options. National Land and Water Resource Audit, Canberra. 2001

[95] Hillel D. Salinity Management for Sustainable Irrigation: Integrating Science, Environment, and Economics [Internet]. Washington, DC: The World Bank; 2000. Available from: http://documents.worldbank.org/curated/en/687661468741583380/pdf/multi-page.pdf [Accessed: 19 February 2020]

[96] Datta KK, De Jong C. Adverse effect of waterlogging and soil salinity on crop and land productivity in northwest region of Haryana, India. Agricultural Water Management. 2002;**57**:223-238. DOI: 10.1016/S0378-3774(02)00058-6

[97] Schoups G, Hopmans JW, Young CA, Vrugt JA, Wallender WW, Tanji KK, et al. Sustainability of irrigated agriculture in the San Joaquin Valley, California. Proceedings of the National Academy of Sciences. 2005, 2005;**102**:15352-15356. DOI: 10.1073/pnas.0507723102

[98] Makarieva AM, Gorshkov VG. Biotic pump of atmospheric moisture as driver of the hydrological cycle on land. Hydrology and earth system sciences discussions. European Geosciences Union. 2007;**11**:1013-1033. DOI: 10.5194/hessd-3-2621-2006

[99] Cook BI, Anchukaitis KJ, Kaplan JO, Puma MJ, Kelley M, Gueyffier D. Pre-Columbian deforestation as an amplifier of drought in Mesoamerica. Geophysical Research Letters. 2012;**39**:16. DOI: 10.1029/2012GL052565

[100] Mcllveen R. Fundamentals of Weather and Climate. 2nd ed. Oxford: Oxford University Press; 2010. p. 632

[101] Bunyard P, Poveda G, Hodnett M, Peña C, Burgos J. Experimental evidence of condensation-driven airflow. Hydrology and Earth System Sciences Discussions. 2015;**12**(10):10921-10974. DOI: 10.5194/hessd-12-10921-2015

[102] Bunyard PP, Hodnett M, Peña C, Burgos-Salcedo JD. Condensation and partial pressure change as a major cause of airflow: Experimental evidence. Dynamis. 2017;**84**(202):92-101. DOI: 10.15446/dyna.v84n202.61253

[103] Sheil D. Forests, atmospheric water and an uncertain future: The new biology of the global water cycle. Forest Ecosystems. 2018;**5**(1):1-22. DOI: 10.1186/s40663-018-0138-y

[104] Lees K, Pitois S, Scott C, Frid C, Mackinson S. Characterizing regime shifts in the marine environment. Fish and Fisheries. 2006;**7**:104-127. DOI: 10.1111/j.1467-2979.2006.00215.x

[105] Carpenter SR, Mooney HA, Agard J, Capistrano D, DeFries RS, Díaz S, et al. Science for managing ecosystem services: Beyond the millennium ecosystem assessment. Proceedings of the National Academy of Sciences. 2009;**106**(5):1305-1312. DOI: 10.1073/pnas.0808772106

[106] Rocha JC, Peterson G, Bodin Ö, Levin S. Cascading regime shifts within and across scales. Science. 2018;**362**(6421):1379-1383. DOI: 10.1126/science.aat7850

[107] Holling CS. Resilience and stability of ecological systems. Annual Review of Ecology and Systematics. 1973;**4**:1-23. DOI: 10.1146/annurev.es.04.110173.000245

[108] Folke C, Carpenter S, Walker B, Scheffer M, Elmqvist T, Gunderson L, et al. Regime shifts, resilience, and biodiversity in ecosystem management. Annual Review of Ecology, Evolution, and Systematics. 2004;**35**:557-581. DOI: 10.1146/annurev.ecolsys.35.021103.105711

[109] Scheffer M, Holmgren M, Brovkin V, Claussen M. Synergy between small-and large-scale feedbacks of vegetation on the water cycle. Global Change Biology. 2005;**11**:1003-1012. DOI: 10.1111/j.1365-2486.2005.00962.x

[110] Rocha JC, Peterson GD, Biggs R. Regime shifts in the Anthropocene: Drivers, risks, and resilience. PLoS One. 2015;**10**(8):e0134639. DOI: 10.1371/journal.pone.0134639

[111] Cvijanovic I, Santer BD, Bonfils C, Lucas DD, Chiang JC, Zimmerman S. Future loss of Arctic Sea-ice cover could drive a substantial decrease

in California's rainfall. Nature Communications. 2017;**8**(1):1-10. DOI: 10.1038/s41467-017-01907-4

[112] Rossi T, Connell SD, Nagelkerken I. The sounds of silence: Regime shifts impoverish marine soundscapes. Landscape Ecology. 2017;**32**(2):239-248. DOI: 10.1007/s10980-016-0439-x

[113] Gordon LJ, Peterson GD, Bennett EM. Agricultural modifications of hydrological flows create ecological surprises. Trends in Ecology & Evolution. 2008;**23**(4):211-219. DOI: 10.1016/j.tree.2007.11.011

[114] Kravčík M, Pokorný J, Kohutiar J, Kováč M, Tóth E. Water for the Recovery of the Climate: A New Water Paradigm. 1st ed. Košice: Typopress Publishing House; 2008. p. 122

Chapter 4

Water Plant and Soil Relation under Stress Situations

Adrijana Filipović

Abstract

Water is an important component in every plant's life helping them to perform basic metabolic processes. The biggest challenge of today's agriculture is how to ensure sufficient water needs at the key phase of plant development and how they can use the available moisture in the soil through the rhizosphere system.

Keywords: water cycle, water properties, water resources, agriculture, water stress

1. Introduction

Water and sun radiations are the most important environmental factors that make life possible on earth [1]. The lack of fresh water is one of the greatest concerns of humankind [2]. The rivers, lakes, reservoirs, creaks, or streams are natural water resources important for every living organism [3]. Without water, the earth would be a dead desert [4]. Water is a prerequisite for life, is involved in almost all processes of life on our planet, and has many functions in the climate system as well [2]. All organisms contain 50–90% water, some aquatic organisms even 99% [3]. If water becomes scarce or has poor quality, plants and animals die. Humans must drink about 2 L of water per day [5]. Any other substance cannot substitute the function of water. Around 97% of water on the earth is salty and unsuitable for drinking and irrigation, whereas 1.8% was frozen in glaciers and snow [6]. About 20% of the world's population already is suffering from water scarcity [3]. Water will be the most important substance during this century, and therefore, we need a global water policy guided by the United Nations [7]. These surface water resources are extremely important on daily basis for human population too. The earth's surface is 71% covered by the oceans regarding the total water sources of 97% [3]. The freshwater resources make up only 2.5% of total water [6]. In addition, 75% of freshwater is made up of glaciers and polar ice that leaves less than 1% of available fresh water in liquid form [8]. Soil water has many roles but some of the most important are water as a solvent, temperature buffer, and metabolite activator [9]. All of these roles are incorporated trough the water-plant-soil relations. During the cultivation of certain plants, many of the farmers found that a number of agronomic measures do not give a good result if the irrigation of crops was omitted or neglected. Water deficit was more often associated with nutrient deficiencies through their reduced solubility and limited distribution to the root system rather than reduction of insufficient nutrient amount. Water needs are increasing on a daily basis that is to be expected compared to population growth; however, water resources are declining and could soon become scarce. Of the total water sources, the most demanding is agriculture and the water used for irrigation for a 68%, while about 21% used for public supply and about 11% accounted for by industry process. Water use around the world has increased six-fold in the past 100 years, twice as fast as the human population, and is expected to double again before 2030, driven mainly by agriculture and irrigation [1]. Water is an important component in every plant's life helping them to obtain their nutrients (through the process of photosynthesis), growth (cell division, mitosis), respiration (cellular respiration), and turgidity (up standing form). Water helps plants to maintain their formation by transporting water and dissolved nutrients, amino acids, and other osmotic active substances from soil to aboveground plant part. Water help plants to perform the most important process, for them, the photosynthesis. In this chapter, we focused in water cycle and its importance for soil to plant life. The soil consists of different horizons, different thicknesses formed under the influence of pedogenetic factors, and processes that have been going on for millions of years and that are constantly going on. Soil system is consisting of solid (soil particles), liquid (water), and gaseous phases (air). A porous space was built between the soil particles of different shape and dimensions in which there is water, air, or some other gas. Vegetable production largely depends on the quality, type, and types of soil, and the necessary factor for plants is water. It is therefore not surprising that all civilizations in the development of human society have settled in the river valleys. Water is constantly present in the soil or on its surface. Its content in the soil is constant changing and depends on weather conditions and the needs of the plant world. The water is in constant circling, and this movement was called the hydrological cycle. Due to the movement of water in the soil and variable content, there are two major problems. One of those problems is excess water in the soil, so due to such water-air regime, unfavorable living conditions for plants occur. Another problem was the lack of water in the soil for normal growth and development of plants that was negatively reflected on yield. The fact is that producers were increasingly faced with prolonged droughts during the growing season. The only measures to combat the consequences of such troubles are the introduction of programs of the irrigation through reclamation measures as a necessary item for the future of agricultural production and reduction of far-reaching consequences if not prepare for the changes that have taken place. The biggest challenge of today's agriculture is how to ensure sufficient water needs for growing crop. This entails to identify critical plant growth phase in order to meet their water needs. Defining the period, form, role, and amount of available moisture for plants could be useful for obtaining the optimal yields.

2. Chemical and physical water properties

The two hydrogen atoms bonded by covalent brigs to an oxygen atom makes water molecule [10]. A water molecule is a polar molecule meaning that it is electro negatively charged (around oxygen atoms) at one end and electro positively charged (about two hydrogen atoms) at the other end [11]. When water molecules are interconnected, the positively charged end of one molecule (hydrogen atom) connected by a hydrogen bond to the negatively charged end (oxygen atom) to another water molecule [11]. The phenomenon of attracting water molecules is called cohesion [10]. Due to its properties (polarity and the formation of hydrogen bonds), water participates in many interactions and plays a major role in the plant organism (because it makes up between 80 and 95% of the mass of plant tissue) [10]. Polarity makes water the most widespread and most important solvent in nature. The polarity property allows the dissolution of ionic substances and organic molecules containing polar groups (OH^-, NH_4^+, and COO^-) [10]. Hydrogen bonds formed between water molecules and ions/polar substances reduce electrostatic interactions between charge-carrying substances and thus allow them to dissolve. The polar parts of the water molecule can form an aqueous mantle (hydration mantle) around the charged particles of the macromolecules, thereby reducing the interactions and binding of the macromolecules, thereby increasing their solubility in water. Due to the large number of hydrogen bonds that connect water molecules, it can absorb heat without large changes in temperature, making it an ideal medium for thermoregulation. As the temperature rises, water molecules movement accelerates. Compared to others liquids, to increase the water, temperature required a relatively large amount of energy (to break hydrogen bonds). From this fact arises one of the basic functions of water in the plant and that is the regulation of plant temperature (thermoregulation) [10]. Since cells contain large amounts of water, they can receive or lose heat with minimal temperature changes. This property can serve to protect organisms from sudden changes in temperature, because organisms that contain larger amounts of water in their tissues were better protected from temperature changes caused by oscillations in heat. Water exists in three different states: ice, liquid, and as steam or water vapor [12]. Water conversion into water vapor requires break down of hydrogen bonds. Therefore, a large amount of energy is used to evaporate the water, which allows the plants to cool efficiently [12]. On the contrary, low temperatures make water molecules approaching. They are closest at 4° C, although they are still moving. At temperatures below 4° C, the molecules just vibrate and the hydrogen bridges become open and rigid [10]. Water in solid state (ice) has a lower density than in liquid, so it floats on the water surface. This property allows many organisms to survive under the frozen water sheet [10]. Water molecules also have a pronounced surface tension. In order to increase the contact area between water and air, it is necessary to break down hydrogen bridges between water molecules [12]. This process requires energy investment, and this energy represents the surface tension. Surface tension has a major role in the transport (movement) of water in the soil-plant system. In addition to surface tension, cohesion and adhesion forces were important in water transport. The attraction between two water molecules is a force called cohesion, while adhesion causes water molecules to adhere to another solid (e.g., a cell wall) [10]. Cohesion, adhesion, and tension allow the appearance of capillarity, that is, the rise of water column through a narrow pipe—capillary, where the water level in the capillary is higher compared to the water level in the source that supplies the capillary [11]. Capillary occurs in various media (soil, root/stem tissue) due to: (1) the attraction of water molecules to the cell wall (adhesion), (2) mutual attraction of water molecules (cohesion), and (3) surface tension of water [13].

The plant tissue consists mostly of water and to a lesser extent of inorganic substances that plants receive from the soil, organic substances formed by photosynthesis and the products of their conversion. Water makes up 80–95% of the mass of metabolically active plant tissue. Majority of the contained water in plant tissue serves for transpiration and less than 1% was used for metabolic activity [14, 15]. High water content is an essential feature of all metabolically active cells. Relative water content reduction up to 70–80% in most plant cells results in the inhibition of central metabolic functions, such as respiratory processes and photosynthesis. In the process of photosynthesis, water is the carrier of electrons and protons. The role of water in plant tissue is multiple. In addition to being the best and most common solvent, water is also a medium for the movement of molecules inside and between cells. It greatly affects the molecular structure and properties of proteins; nucleic acids and other macromolecules indirectly affect the property of cells plasma membranes. Water is the basic constituent of protoplasm.

Water is essential for most of biochemical reactions in plant cells. Water is directly involved in number of chemical reactions, for example, hidrolysis and dehydratation reaction (e.g., ADP phosphorylation is actually dehydration of the

ADP molecule, is a donor of hydrogen in photosynthesis). In addition, as water has a high heat capacity, its presence in the plant ensures that temperature changes occur slowly. The polarity and ability to form hydrogen bonds allows water to participate in a number of interactions. Water molecules were arranged around ions or charged groups of macromolecules and cover their charge. This reduces the interactions between the charged substances and increases their solubility. Therefore, water is the best solvent for ionic substances. Water is in liquid phase and medium in which all enzymatic reactions take place. Various substances can enter into a chemical reaction with each other only if they dissolve in water. The uptake of solutes is possible only from an aqueous solution. As the water content in the plant organism decreases, so does the vital activity. Water regulates turgor pressure and upright visual appearance and cell size. A large number of metabolic functions of water were realized by the processes of uptake and release (transpiration).

3. Water uptake and movement trough the plant

3.1 Basic process of water uptake

Plants constantly receive water by root system and excrete water by vegetative plant's parts. The root system consists of primary and secondary roots overgrown with root hairs. Root hairs develop on the root surface. Due to their large number, they significantly increase the root area and thus make it easier for the root to absorb water and minerals from the soil. A close contact of the root surface and soil is necessary for the absorption of water by the root. The root hairs penetrate between the soil particles and immerse themselves in the capillary spaces of the soil where the water is located. Under favorable conditions of growth and development, plants scatter the root network and increase its volume. In restrictive plant growth conditions, plant root growth is often primarily restricted. Roots hairs were the most susceptible to insufficient water and nutrient demands resulting in decay, especially within the herbaceous plants [16]. The pathway plant-soil-atmosphere water movement occurs through different media (cell walls, double phospholipids layer, cytoplasm, etc.), and the transmission mechanism changes depending on the type of medium through which it passes (**Figure 1**).

Basic processes that enable the water uptake and conduction of water in the plants are swelling and osmosis. Both processes were conducted by the reduction

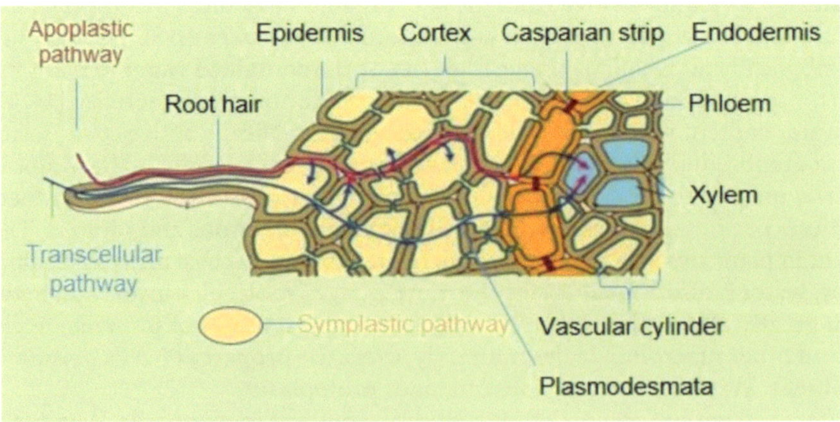

Figure 1.
Water movement and different pathway of water uptaking in roots.

of water chemical potential and by the process of diffusion. Diffusion process presents water movement from the medium of high-water potential to the medium of low water potential. In an aqueous solution of a substance, the gradient of the water chemical potential is opposite to the gradient of the electrochemical potential of the solute, so water diffuses in the opposite direction from the solute. During this process, there is no chemical reaction. The diffusion rate presents amount of substance that diffuses in certain period proportional to the concentration gradient. First Fick's law defines diffusion rate of solvent across the membrane. The process of diffusion is important for small molecules in aqueous solution effective on cell level as for entering of solution in root cell or for a stomata transpiration rate. Diffusion is not effective enough to transmit solutes to the too long pathway. The rate of diffusion is rapid over short distance but extremely slow over long distance. The process of diffusion has great importance in receiving water from the soil, the movement of solutes over short distances, and the loss of gaseous water from the vegetative plant's part, but it is not important in the transfer of water over long distances that occur in the main stream. The cohesion-tension theory explains the mechanism of water transport without consuming metabolic energy. Xylem water transport is closely related to cohesion (water molecules bind to each other) and adhesion forces (where water molecules adhere strongly to the conductive elements of the xylem), forming a transpiration column for which upward movement is responsible for the negative hydrostatic pressure [11]. Unlike diffusion driven by a difference in concentration (concentration gradient), mass bulk or mass flow represents water molecule mass flow often driven by differences in pressure (pressure gradient). This mass flow mechanism by xylem elements frequently is used for water transport over the long-distance pathway.

Mass or free flow of water plays a significant role for absorption of nutrients from soil solution of high concentrations even when transpiration is high. Then significant amounts of water move toward root carrying with it dissolved substances (nutrients) which the plant receives. Certainly, if there is not enough water in the soil, there is no flow of nutrients, while too much water can increase leaching of nutrients or asphyxiation of root.

The root hairs penetrate into spaces between soil particles and allow water to enter the apoplasmic space of the root cells. To receive water from the soil, corresponding reduction of water potential between the soil and roots has to occur. The drier the soil, the more negative the water potential becomes, and the potential swelling pressure increases, as moisture is still retained only between the soil capillaries. The water enters root hairs by imbibitions and osmotic transporting into the central cylinder of the root. Further, the plant uses water potential reduction between the soil and the atmosphere to conduct water from the soil through the plant body to the atmosphere without consuming energy. A process leading to water transport in plant cells is osmosis referring to movement of a solvent such water and other substances across a membrane. The membrane for all living cells presents certain barriers: they separate different parts of the cells from each other and greatly impede movement of substances between compartments. The plant cells membranes are semi-permeable, since they are well permeable to water molecules and other smaller particles with a weakly charged charge and quite limited permeability to larger molecules and particles with a pronounced charge [5]. Osmosis similarly to the diffusion and mass flow occurs and it appears unpredictably as a response to the suction forces.

Water is a key factor in initiating the germination of dry seeds when they reach the soil. The water in the seed stimulates the swelling process by imbibitions; hydrolytic enzymes ware activated; and the germination process begins. The entry of water into the seed is wrapped from the hardness of the seed coat, and sometimes with tightly closed seeds, a negative pressure of −100 MPA [17] can occur when entering the seed coat. Activated hydrolytic enzymes also activate other biochemical processes on which the germination process depends.

Water uptake by seeds is a process called imbibition, after which a certain period causes swelling and rupture of the seed coat. Germinating seeds use reserve nutrients stored in the seed endosperm. This nutrient reserve ensures the growth of the embryo. When the seed absorbs water, hydrolysis enzymes are activated that break down these stored reserve substances into metabolically useful chemicals. After the germ emerges from the seed layer and begins to grow roots and leaves, food supplies are usually depleted; in this case, photosynthesis provides the energy needed for further seedling growth, which now requires a continuous supply of water, nutrients, and light.

The swelling degree depends on the balance between the soil solution water potential and water potential of seeds, described as acting as repulsive and attractive forces between the two charges [18]. Imbibition is a physical process in which water enters the seed coat, where the volume of the seed changes significantly, exceeding its actual seed surface [19].

3.2 Water chemical potential

The term water potential is the most important factor for understanding the way of water moves in plant cells and through the conducted elements of the plant [20]. The chemical potential represents the water potential when it comes to water. The chemical potential of water or any substance is a measure of the available energy per mole by which that substance will react or move. Because the water molecule is neutral, the electrical potential has no effect on the chemical potential of the water. The chemical potential of water is a relative quantity and expressed as the difference between the potencies of a substance under certain conditions and the potential of that same substance under standard conditions. The unit for chemical potential is the energy per mole of a substance (J mol^{-1}) but for better understanding, we use term water potential. The unit of water potential presents the free energy per unit volume of water solution in relation to the standard state of water, as a result of the combined action of solution concentration, pressure, and gravity in ambient temperature regime [19]. Presented equation express water potential as:

$$\Psi = \mu_{w1} - \mu_{w2} / V_m \quad (1)$$

Ψ, water potential; μ_{w1}, chemical potential of water; μ_{w2}, chemical potential of pure water; Vm, molar volume of water.

The molar volume of water presents the molar mass (w) divided by the mass density (ρ) expressed cubic meter per mole (m^3 mol^{-1}). The soil particles and plant tissue cells absorb water, and the level of absorbed water depends on different factors. Absorbed water depends on soil pore size, water regime, ambient temperature, and pressure, and working adsorption forces depend on concentration of solutes in water. Uncontrolled movement and movement speed of some solutes in water depends on their concentration. Restricted movement of water molecules reduce chemical potential of water expressed as the Brownian's irregular water movement. A plant cell does not have mechanism of "water pumps" for water potential incensement so the water uses gradient water potential reduction. To determine the value of the water potential, equilibrium methods are most often used in which the plant tissue is brought into balance with solutions of known water potential. Water potential of pure water is always equal to zero but water in natural system, in plant cells, or soil pores has always dissolved some solutes as organic molecules, sugars, enzymes, or other substances making water potential more negative [19].

This formula describes the energy state of water as the sum of the solute potential, the pressure potential, and the gravitation potential in a mixture of water and other particles in relation to standard conditions. The main factors contributing to the water potential can be expressed by the following equation:

$$\Psi = \Psi_p + \Psi_s + \Psi_m + \Psi_g \tag{2}$$

Ψ, water potential; Ψ_p, pressure potential; Ψ_s, solute potential; Ψ_m, matrix potential; Ψ_g, gravitation potential.

The pressure potential expresses the effects of pressure within the water potential of solution and refers to the hydrostatic pressure. Water potential has elevated by positive pressure, whereas negative pressure has an opposite effect. Plant cells have solid cell walls and can produce strong positive hydrostatic pressure called turgor. In xylem elements and aplastic space, negative hydrostatic pressure can develop which is important for water distance moving through the vascular tissue.

The solute potential expresses the effects of dissolve solutes within the water potential of solution and refers to the osmotic potential. Dissolved substances dilute water and therefore reduce its free energy. On the contrary, diluting a solution with water leads to a decrease in the concentration potential of dissolved particles and to an increase in the concentration potential of water. The effect of the osmotic potential on the water potential of the solution is negative, which means that the solutes reduce the water concentration and it is potential. The values of the osmotic potential and the osmotic pressure differ only in sign. In other words, the values of the osmotic potential are negative, and the values of the osmotic pressure are positive.

The gravity potential (Ψ_g) represents the gravity effects on water potential of some solution. Gravity causes the water to move downward, and the potential affects the movement of the water depending on the height or transport distance. The effect of gravity on the potential can be neglected in the case of the transfer of solutes between cells or when the substances transported at a shorter height than 5 m.

One of the important components of water potential is potential of matrix (Ψ_m). This parameter is important for reducing the water potential because of the action of water on a solid surface such as a cell wall or soil particles. Such interactions of water and solid reduce the tendency of water molecules to chemically react or evaporate. In addition to dissolved particles and colloidal dissolved macromolecules, the cell has membrane surfaces and hydrated structural elements that can affect the water potential. For example, the water potential of a cell wall is not equal to zero even when the solution in the cell is pure water. The contribution of the cell wall structure to the water potential consists of a negative pressure component caused by water bound in the capillaries and an osmotic component that can theoretically be included in both the osmotic and turgor potential. For a practical reason, these potentials were often combined into one called the matrix potential. In an adult vacuolated cell, the matrix potential of the protoplasm and cell wall is small compared to the osmotic potential that was often neglected. The more negative the water potential of a cell, the higher its suction force. In water-saturated cells, the water potential is equal to zero, and the turgor potential corresponds to the sum of the osmotic and matrix potentials. The water potential increases with increasing turgor pressure and decreases with increasing osmotic pressure. The values of turgor and osmotic pressure depend on temperature and increase with increasing temperature. Water moves exclusively from the area of higher water potential to the area of lower water potential, that is, down the chemical gradient. On the contrary, water from the area of lower water potential can be transferred to higher area only with energy consumption (endogenous process). Water will enter the cell until the water potential inside the cell equals that outside the cell. The flow of water into the cell resulting from the water potential gradient causes a hydrostatic pressure (turgor) in the vacuole. This pressure gives the cell tension and strength. Since the cell wall is quite solid but also elastic, small changes in cell volume can cause large changes in turgor pressure.

The turgor pressure is very important for the upright appearance of the plant and the strength of the cells. In conditions when the plant loses water, due to the limited availability of moisture, the turgor pressure decreases, the cell walls relax, and the plants take on a withered appearance. The turgor pressure acts against further osmotic flow of water in the vacuole, as the actual operating pressure for osmotic flow and at disposal, it has only a part of the potential osmotic pressure that was not compensated by turgor and was called the tension or suction force. The positive value of the suction force corresponds to the negative value of the water potential. The difference in water potential between the outer and inner membrane space is the force that allows water to be transported osmotically. The potential osmotic pressure of a cell decreases due to water intake and increases with water excretion. The cell sap contains a relatively high concentration of solutes and entering of water molecule trough the cell wall has controlled by pressure. The turgid cells have a suction force equal to zero, whereas turgor and osmotic pressures were equalized. Plant cells between the saturated state and the wilting state have a suction force corresponding to the negative value of the water potential. When the plant cells are partially dehydrated, the turgor pressure corresponds to zero and the suction force corresponds to the value of the potential osmotic pressure. When the plant cell is saturated, the turgor pressure takes a negative sign, and the suction force corresponds to the sum of the turgor and osmotic pressure. Small changes in Ψs usually accompanied such changes of turgor pressure. The water uptake to the cells leads to positive sign of pressure potential Ψ_p. Water absorption through the roots is possible only when there is a corresponding drop in water potential. The water potential was significantly affected by humidity, because in conditions of high humidity, the water potential is quite high, while in conditions when the air is quite dry, there are large differences between aboveground plant parts and the atmosphere. The amount of water available to the plant is in balance with the physiological capacity of the plant to absorb water and the environmental conditions that affect the intensity of plant transpiration. The movement of water in the plant-environment system always takes the place of water and solutes movement from the area of higher potential to the area of lower potential until the concentrations equalize. The water potential in the leaves of plants rating from –10 to –100 bar, while in the atmosphere, in the conditions of relative humidity, up to 50% of the potential reaches up to –1000 bar. A large difference in the water potential gradient of plant leaves and the atmosphere causes transpiration. Negative water potential usually occurs in leaves than the root cells. Similar happens when water enters the root of plants, through the apoplast where the driving force is the difference between the water potential of the xylem and the soil solution. When the water potential of a xylem solution is more negative than the water potential of soil solution, water enters the root and moves all the way to the endoderm. Van den Honert [21] explains the differences in potential that occurs when water enters into the root or exits by transpiration. An increase in the water potential due to an increase in relative humidity of air and a decrease in the water potential of the soil due to desiccation leads to a slower transfer of water from the roots to the aboveground organs of the plant. The lack of water is most reflected in the decline in turgidity of plant cells and thus in the overall appearance of the plant, with the leaves falling down, twisting, decreasing the intensity of photosynthesis, and reducing all other metabolic activities [22]. Any deviation of the available amount of moisture from the optimal in plants results in stress and the plant wilting and earlier the deterioration of plant tissue occurs [23]. In conditions when the plants are short of moisture for a long time, it passes into the phase of permanent or reversible wilting and then dies.

3.3 Plant water balance water status

Plant water balance water status in plants is very important for plant growth and development, and metabolic activities, and especially, the lack of water drastically affects the length of the growing season and crop yields [23]. The total ratio between the water absorbed by the root system and the water released by transpiration from aboveground plant part is referred to as water balance. In cases when the transpiration of water exceeds its absorption from the substrate, water deficit occurs. The physiological processes in a plant depend on the amount of available moisture. The lack of water leads to inhibition of growth and photosynthesis. In addition, it acts on cell division and inhibits the synthesis of cell wall components and proteins, causing a stomata closure. Environmental conditions significantly affect the water potential, which is negative in conditions of well-available humidity, and in arid environments, the negative pressure was even more pronounced. Water potential was not only response to the environmental condition but also to the plant genetic characteristics [19]. The water transfer presents passive process so water movement to the plants occurs due to the low water potential of the plant in regard to the water potential of the soil. Since turgor pressure cannot be directly affected, the plant cell can regulate the water balance only by active regulation of the osmotic potential. Osmoregulation is the adaptation of osmotically active substances in the cell to the newly occurring environmental changes that largely control the water potential of the cell. This adjustment can be achieved by receiving/releasing or synthesis/decomposition of osmotically active substances. Under drought stress, plants are able to maintain water adsorption by increasing the cellular solute concentration, a process called osmotic adjustment. Osmotic adjustment occurs when the osmotic potential of a cell changes due to an increase or decrease in the content of osmotically active substances. Many cells respond to water stress by increasing the water potential of the cell. In this way, the decline in turgor can be prevented or minimized. Osmotic adaptation is an important feature of delaying dehydration in water-constrained environments because it maintains cell turgor and physiological processes with the development of water deficit [24]. Osmoregulation and osmotic adjustment are two different mechanisms. Glycophytes and halophytes have the ability to adapt to stress in two ways. One is that under conditions of water deficiency, plants can synthesize organic substances such as organic soluble substances, such as glycine betaine or proline, or high concentrations of inorganic ions [25]. The water balance of a plant depends on the rate of water uptake and excretion.

3.4 Absorption, transport, and transpiration of water

Water movement in the plant occurs predominantly regarding the passive or active transport of osmotic active substances along with water across the membrane. Main trigger for such solutes moving through the plant cells is difference in water potential followed by a difference in pressure (**Figure 2**).

The entry of water into the root cells occurs passively, that is, diffusely, and the solution moves freely through the apoplasmic space. Since the water molecule uncharged, it can very easily cross the membrane without hindrance and continue to

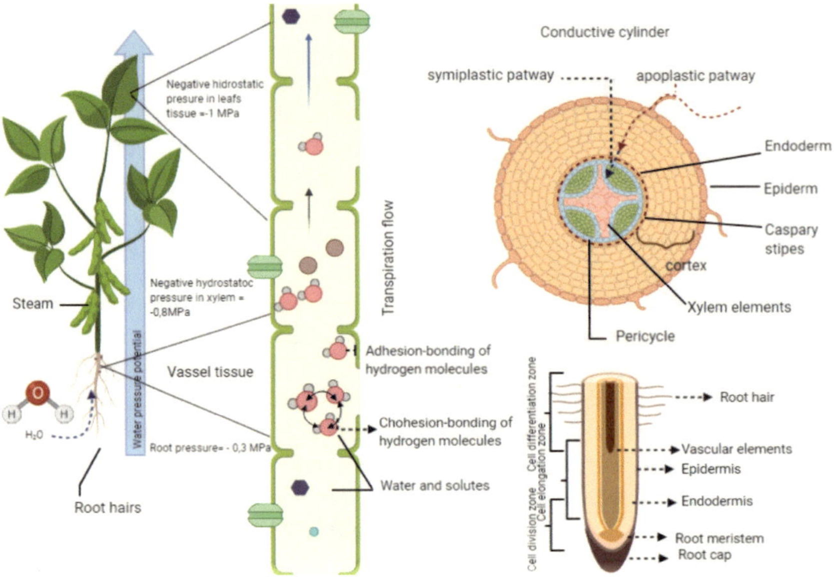

Figure 2.
Water movement through the plant root and root elements.

move upward through the conductive elements of the xylem by mass flow. Mass flow is also a passive mode of water transport, which is used for long-distance transport. While diffusion is a way of moving water and solutes over shorter distances which mainly occurs at the entry of water into the root cells and the exit of water through the stoma into the atmosphere, which occurs mainly in nonvascular tissues [26]. In the rhizosphere layer, water generally moves by mass flow to the site of adsorption. However, after contact of water and solutes with the root hairs, the mode of uptake changes significantly as other forces occur that affect the uptake mechanism. Water movement trough the rhizosphere layer depends on the texture and structure of the soil. Since more permeable, sandy soils have weaker buffering capacity, so they tend to dry out quickly, while compacted, clayey soils have very limited capacity to receive and conduct water and nutrients, and transitional soil types are in terms of permeability and moisture retention and nutrients mobility of moderate capacity.

3.4.1 Absorption of water

Absorption of water requires a close contact of the roots of an intact plant and soil particles in the aqueous soil phase. The larger the volume of the roots and the root zone in the rhizosphere layer of the soil the greater the possibility of absorption. Root hair presents a tissue of the rhizoderm (root epidermis) that pronounce the root surface area affecting the root capacity to absorb water and minerals. Water enters the root in the root hair growth zone and in the root tip zone. Older parts of the root are often impermeable to water. Cracks in the root bark, as well as the growth of the secondary (lateral) root, allow water to be received by older parts of the root as well. Plants have relatively low water use efficiency and therefore need to receive large amounts of water. The water potential reduction among soil particles and the root hair occurs in order to root absorb the moisture from the soil. Water potentials in the soil (expressed as the suction tension in the soil) determine the potential imbibitions pressure caused by hydration and capillary forces. Water potential becomes more negative as soil dries out, and the imbibitions pressure increases potential, because water retained only in narrow capillaries, which is also case with potential osmotic pressure. Root growth usually follows soil moisture sources and growth intensively while those parts that are no longer actively involved in absorption die off, resulting in asymmetric root growth.

3.4.2 Water transport

Water can move from the epidermis to the root endoderm in such a way that in apoplastic cells, water passes exclusively through cell walls and intercellular spaces, without passing through the membrane, while in the cell pathway, water passes through protoplasts. The cell pathway has two components: the transmembrane pathway in which water passes from cell to cell through plasma membranes and the symplasmic pathway through which water from cell to cell passes through plasmodesmata [11]. When passing water to the root endoderm, all three types of pathways were usually combined. The root endoderm is single layered and separates the conductive cylinder from the root cortex.

The movement of water through the apoplast of the endoderm of the cortex blocks the Casparian stripes located in the radial cells of the endoderm. Casparian cells are narrower or wider suberized cells, located around vascular elements of plant tissue. In younger cells, these stripes are similar in thickness to other cell sections, while older cells show these sites as more pronounced. Casparian stripes are barriers, so the water flow and solutes at that point cannot pass through the intercellular space of the root parenchyma, but must take place through the plasma membrane all the way to the endodermal cells—*apoplastic pathway of water movement*.

The *symplastic pathway* of water movement takes place from cell to cell through plasmodesmata where water does not cross cell membranes. Unlike the apoplast, the symplast is the living part plants composed of interconnected cytoplasmic cells with the help of plasmodesmata that grow on the cell wall interconnecting cells into a symplast. When water moves through the apoplast and symplast, water does not cross the cell membrane, because this water movement caused by difference in hydrostatic potential.

Water channels as the integrated part of membranes mostly responsible for transcellular transport of water, of which aquaporin's standout, which originate from larger protein families of major intrinsic proteins (MIP) forming pore channels in cell membranes and mediate in many other physiological processes [27]. The transcellular pathway is the movement of water from cell to cell where water crosses cell membranes, entering and getting out of the cells. The transcellular pathway also includes the entry of water into the vacuole, that is, the transport of water through the tonoplast. Because of that, the transcellular water movement was driven by the difference in gradient of water potential.

Xylem water transport presents the longest path taken by water in a plant (long distance transport). Almost total amount of water moved through the vascular cells was transported by xylem. Compared to transporting water through living cells, the xylem built of dead cell that provides little resistance to water movement. Xylem presents specialized tissue cells for the water and solutes transport. The xylem elements anatomy allows very efficient transport of large amounts of water for long distance. The two types of dead cells that make up the xylem elements: tracheid and trachea. These are cells with lignified, thickened secondary cell walls. Tracheids are spindle cells that communicate with neighboring cells through numerous pores in the walls. Pores are microscopic areas in which there is no secondary wall, and the primary wall is thin and porous. The tracheae were shorter and wider than the tracheid and have perforated walls that form perforated plates at the end. The conduction of water by xylem elements is the result of the action of cohesion and adhesion forces on the conductive wall elements. Within the conductive elements of the xylem, continuous water columns are formed, which can, due to high tension, cause the interruption and appearance of air bubbles or the so-called embolism. The xylem transport presents of water movement for a long distance pathway from soil to aboveground vegetative plants part under the differences in pressure gradient [28]. The pressure gradient is responsible for the main transpiration flow of water, however, sometimes although less efficient transpiration of water can take over the root pressure in certain condition. In that condition, root pressure mostly show values less than 0.1 MPA [29] that correspond very high humidity as result of the high difference between daily and night temperatures. In such terms, guttation could occur as a result of water transport on leaf edges through specialized pores called hydathode [30]. In addition to guttation, root pressure is also responsible for the appearance of exudates of plant tissue in the case of mechanical injuries or cuts [31]. The amount of exudate that the plant secretes primarily depends on the condition of the plant and environment. Root pressure consumes metabolic energy and can therefore be inhibited by respiratory toxins or low temperatures in the root area. Water movement through the xylem requires the differences in gradient pressure; it can also occur due to the negative pressure (vacuum) that develops by transpiration (loss of water through the coup). The root pressure (0.05–0.5 MPa) cannot develop in conditions of low soil water potential or intensive transpiration and given that the pressure required for long-distance water transport is up to 3 MPa. Due to the loss of water by transpiration, tension (negative hydrostatic pressure) occurs which moves the water upward in xylem. For the transport of water and solutes from the roots to the aboveground parts, the most significant suction force is the aboveground organs of the plant, which create a negative pressure as a result of transpiration and root pressure. The mechanism by which tension drives water by xylem called transpiration suction.

3.4.3 Transpiration of water

Negative hydrostatic pressure develops on the surface of the cell walls below the stomata due to the loss of water by transpiration. The water movement in vascular elements of plant tissue was explained by transpiration cohesion tension model due to the forces that act based on the existence of soil-plant atmosphere continuum. Transpiration makes releases of water through the stomata on leaf surfaces. The capillary and cohesive forces cause water to enter the root cells from the soil, then go through the xylem elements all the way to the leaves of the plant to make up for the lost water. Xylem is a passive conductive element through which water columns move under pressure. Therefore, on the upper side of the water column, the water is sucked due to transpiration, which allows the entire column to move upward. In order to increase the contact area between water and air, it is necessary to break the hydrogen bonds between water molecules, for which energy must be invested, and this energy represents the surface tension. Surface tension plays an important role in the transport (movement) of water in the soil-plant-atmosphere system. In addition to surface tension, cohesion and adhesion forces are important in water transport. Cohesion is the force by which water molecules are attracted to each other, while adhesion causes water molecules to adhere to another solid (e.g., a cell wall). Cohesion, adhesion, and tension allow the appearance of capillarity, that is, the rise of a column of water through a narrow tube—capillary, where the water level in the capillary is higher compared to the water level in the source that supplies the capillary. Adhesion and surface tension together pull the water column in the capillary allowing it to move upward. The height of the water column depends on its mass, and the water column will rise through the capillary until the mass of the water column equals the action of surface tension and adhesion forces. The narrower the capillary, the higher the water column in it will be because it will have less mass and the adhesion and surface tension will be higher.

The leaf cell wall system acts as a network of microscopic pores filled with water that adheres to cellulose microfibrils of the walls. Leaf mesophilic cells are located below the stomata, and in conditions when the stomas are open, they are in direct contact with the surrounding atmosphere. As water evaporates into the surrounding atmosphere, the surface of the water retreats into the interspaces between the cells where curves of the contact surface between air and water are created. Because of the surface water tension, menisci cause tension, that is, negative hydrostatic pressure. As water evaporates more and more, the menisci become deeper (more curved) and the tension increases (more and more negative hydrostatic pressure).

By moving water from the root to the stem through the xylem, water enters the leaves via the petiole. The petiole xylem redistributes water to the edge of the main leaf vessel, which then branches into progressively smaller veins and is incorporated into the leaf mesophyll. Since different plant species have different anatomical leaf structure and thus the arrangement of veins on the leaf for dicotyledonous plants, it is considered that most of the water used for transpiration of this plant is stored in smaller veins [32, 33]. After the water leaves the xylem, it moves through the bundle cells that surround the veins. It is not yet clear which exact path of water follows after exiting the xylem through bundle cells and enters into mesophilic cells, but the apoplastic pathway during transpiration probably dominates [32].

Xylem elements are structurally adapted to large changes in pressure. Pressure changes are dependent on temperature oscillations and, as mentioned earlier, can cause bubbles to appear in the conductive elements—embolism or cavitation. Such bubbles only briefly interfere with transpiration flow, and this problem easily overcomes, thanks to the numerous pores in the walls of the trachea and tracheid. Such occurrences of bubbles in the conductive elements interfere with the normal transpiration flow and the established pressure, so they can often affect the photosynthetic activity and other physiological processes of the intact plant [34]. Embolic or cavitation is often occurred in very high tree that can develop high tension needed for transpiration flow. The cell wall balances the volume changes without major change in water potential that occurs because of water loss by transpiration. Changes in water potential are related to changes in volume in cells that are mainly the result of transpiration. Differences in hydrostatic potential (Ψp) result in changes in cellular water potential (Ψw) with respect to the cell wall strength. On the way out of the leaf to the atmosphere, water passes from the xylem to the cellular walls of mesophilic cells and from there evaporates into intercellular spaces of leaves. From the leaves, water released by diffusion in the form of water vapor through the small openings of the dental apparatus (stomata), which are located in the epidermis of the leaf, a process that is called transpiration.

The actual flow rate of the xylem content is difficult to determine because the substances traveling through the xylem were constantly alternating with the environment of the conductive elements. The rate of transpiration flow increases with increasing transpiration rate until rupture control begins to operate, and there are no difficulties in water supply. In this case, short-term fluctuations in transpiration can be manifested as changes in the speed of transpiration flow. The speed of the transpiration flow shows the daily rhythm: in the morning, the transpiration begins and the water movement starts, and in the evening when the stoma close, the transpiration flow is interrupted. This process does not require energy and it is a passive way of water flow. The plants transpire most of the water over the stoma and release the water in the form of water vapor. Some of the moisture can be evaporated by plants through lenticels or cuticles, although it is a very small amount of moisture. The intensity of transpiration is related to the size of the leaf area, the number and size of stoma, the appearance of the leaf surface, and of course, the environmental conditions in which the plant is located. If the plant suffers more damage during intensive transpiration, it must compensate for the water by receiving it from the soil. The numerous open stomata allow the exchange of O_2 and CO_2 gases and the evaporation of water. If the air immediately around the leaf is dry, water vapor molecules move from saturated air to the unsaturated external atmosphere according to the law of diffusion. The function of the stoma is to facilitate the excretion of water vapor by opening it, and on the other hand, to make it difficult for stomata transpiration by closing it with the insufficient water supply [35]. The mechanism of opening and closing of the stomata works based on the water and osmotic potential of the gate cells. In order to act by opening of stomata, the gate cells after the water entering should have water potential lower than the water potential of the surrounding cells. The water potential of the gate cells largely depends on the osmotic potential. Since gate cells contain chloroplasts, they also show photosynthetic activity. The leaves under daily conditions where maximum transpiration occurs release 50–70% water vapor. In the light period, intensity of photosynthesis may decrease for 50% or more due to the limited water supply [36, 37]. Since the gate cells are photosynthetically active, this enable them the accumulation of sugars, which reflected the osmotic potential, and the regulation of turgor pressure, responding on the mechanism of closing and opening the stomata [38]. Water has important role in the mechanism of opening and closing of stomata. When the plants are well supplied with water, the guard cells are turgescent and the stomates are open, while in conditions of water deficit, the guard cells lose turgor and the stomata are closed. Model of opening and closing of stomata would be used in genetic engineering for producing of species with reduced water requirements and better production rate [35]. The physiological activity of plants is significantly disrupted by interfering with the process of photosynthesis either through a process of reduced transpiration or altered gas uptake and release [39]. This is one of the reasons of balance maintaining between process of photosynthesis and transpiration. Concept soil-plant-atmosphere continuum is based on the decrement of tension of sap flow through the vessels, and transpiration flux is proportional to the pressure gradient in leaves [40]. Transpiration into cormophytes mostly shows

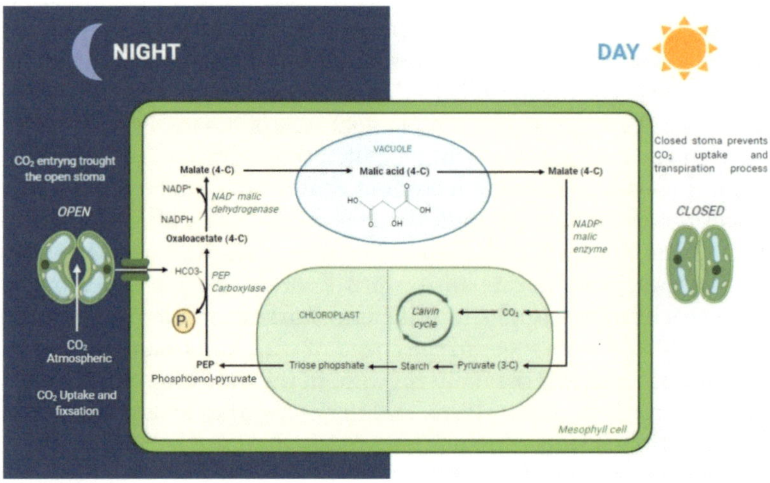

Figure 3.
CAM—Crassulacean acid metabolism in plant cells of arid area.

a characteristic diurnal rhythm. Time control of the opening and closing of the stomata serves to maximize photosynthesis and minimize transpiration. At night when there is no photosynthesis, there is no need for CO_2 absorption either, the stomates are closed and unnecessary water loss is prevented. While in the morning, when the water supply is abundant and the sun's radiation is conducive to strong photosynthetic activity, CO_2 requirements are pronounced and stomates are open. Some succulent's plant have crassulacean acid metabolism (CAM) that enables plant to keep stoma open during the night and uptaking the CO_2 and water making the acidification process through the malic acid building up in vacuole (**Figure 3**). During the day, stomas are close and transpiration as well as CO_2 fixation stopped, and in Calvin process, starch compounds were created.

4. Plant adaptation to the soil moisture regime

Soil moisture is primarily important for water circulation in continuum soil-plant-atmosphere system. The importance of moisture is especially emphasized for the life of terrestrial plants, which are tied to the soil by their roots. Plant roots from the soil absorb water and solutes, so that they can grow and develop. The water that the plants absorb passes through the conductive elements of the plant, reaches the vegetative parts, and then comes out as water vapor trough the stomata. We say for this process that the plant transpires. The needs of plants for water vary, but the fact remains that water in agricultural production is one of the main limiting resources in gaining the optimal yields, with the implementation of regular agro-technical measures. Soil is supplied with moisture through precipitation; however, soil moisture and moisture retention in soil pores and the pathways by which it reaches plants are different depending on the buffering capabilities of the soil. It is clear that some plants remain viable even after a long drought, because their roots manage to find moisture in the deeper layers of the soil. There are also those plants that, due to their anatomical structure, have adapted to life in arid environments. This mode of survival allows them to have large water storage capacities in meso-philic leaf cells as they have succulent leaves. Their leaves transpire in very limited quantities of water, because otherwise they would die very quickly. There are thousands of different species of plants that are adapted to living in desert conditions. Among them are common plants that are classified as succulents. The term succulent can be explained from different perspectives but is most commonly used in terms of a plant that has specialized tissues for water storage, resulting in their special morphological characteristics: thick, fleshy stems, leaves, and/or roots. Sometimes the leaves are transformed into thorns, with the photosynthetic function taken over by a thickened green stem, in other cases by geophytes that have most of their thickened water storage tissue underground, and the third is large trees that store water in huge swollen trunks. Of course, there is a continuum among plants from those that have almost no water storage tissues to those that possess highly developed tissues intended for that purpose so it is difficult, or even impossible and inaccurate, to speak of plants that are and are not succulents. It might be more accurate to use the term "plants with pronounced succulent characteristics," but for simplicity, the term "succulent" is used. Terrestrial plants usually adapt to moisture conditions and soil type by developing roots that provide them with water and nutrients. There are plants that grow on soil with very little water supply and where the plant faces drought for most of the year. In such plants, the roots are often adapted in such a way that they exceed the volume of the aboveground part of the plant. In desire, they stretched and branched to the extent that there was favorable moisture content in the soil, from which it draws strength for its growth. Often such plants show both weak growth and progression but manage to survive dry periods. Because water in deserts does not stay in one place for long and often drains very quickly, without roots penetrating deep into the soil succulents depend on the root network close to the surface (the first 5–40 cm below the surface) to collect most of the water after rain. Therefore, many species of cactus, agaves, and other succulents' groups have many shallow and extremely widespread roots, which is an adaptation to rains that are sudden and short-lived. Within a few hours after the rain comes, many additional fast-growing lateral roots appear that have thin walls and increase water absorption. When water disappears, they degrade [41, 42]. Often, as an example of a drought-adapted plant, one specific plant, the Jericho rose or *Anastatica hierochuntica*, is found in the list. In addition, succulents were defined by their morphological characteristics, it is important to define them from an eco-physiological point of view: succulents plants have ability to survive in a water-restricted habitat using special strategies for water use. Succulents can be in different life forms (annuals or perennials, shrubs, and trees) and from completely different genera and families.

Some economically important plants do not have the ability to adapt to drought (**Figure 4**), such as the aforementioned wild plants of arid areas. In complete lack of water or prolonged drought, they dry out completely and die. In addition, during periods of water deficits, these plants show a completely different course of development, in a way that they often discard their leaves and fruits and slow down growth.

Therefore, it is extremely important that plants receive adequate amounts of moisture during growth and development in order to be able to optimize the level of expected yield. Moisture from the soil is a much more favorable factor that affects the uptake and utilization of water in the plant compared to the moisture that the plant receives through precipitation [42, 43]. The biological importance of precipitation for the plant itself is questioned, because short-term rainfall does not have high efficiency, unlike longer weather conditions or sudden precipitation,

Figure 4.
Economically important crop Vitis vinifera L. on plantation under drip irrigation (left: Cover crop system with irrigation) and individual farm with open filed crop growing system (right: Consequences of drought).

Figure 5.
Soil desiccation and cracking (left), aromatic plant root adaptation to arid land (middle), and very fertile sinkholes supplied by runoff water (right).

and in terms of their effectiveness, same authors have discussed, the distribution of precipitation during the growing season when the plant has the greatest moisture needs is questioned [43]. Meaning that distribution of rainfall over the vegetation season is a key factor in plant productivity because of variability in plant phenology requires a suitable frequency of precipitation periods rather than suitable total amount during the total year (**Figure 5**).

5. Soil moisture sources

Soil supplied with moisture from the atmosphere and/or from deeper soil layers. The precipitation from atmosphere could be in the form of rainfall, snow, ice rain of some other forms in which riches the soil. From the deeper layers of the soil, the earth was supplied from groundwater. Groundwater was created by the underground discharge of water from precipitation from the hills to the lowlands. Sometimes these waters can appear on the surface, which we call natural springs. Groundwater varies in its depth. The one located 1–2 m below the soil surface is very useful for plants because it protects them from drought. The agroclimatology as a science pays more and more attention to temporal and spatial variations of moisture as an essential element in the surface distribution of moisture and energy source that are reflected in the complete ecosystem influencing the soil-plant-atmosphere continuum [44].

All soils are water permeable, because water can move through the space of interconnected pores between solid particles. Soil moisture behavior due to gravity and other factors such as moisture quantity, distribution, and moisture pressure significantly affect soil properties. In conditions when the soil was saturated with moisture, all soil pores are filled with water, and gases are expelled, which often results in anoxia in plants, especially if the water overloading conditions are prolonged. If not, all pores ware filled with water; the soil is partially saturated or unsaturated. The highest plant yields are achieved when the most favorable ratio of air and water in the soil was achieved and especially in the critical periods of each crop [45]. Since different plant species may have different water needs, which also depend on the developmental stages of each plant, it is necessary to provide the required amount of water in critical periods of plant [44]. Climatic characteristics and soil water regime and their interrelationship is hard to have in balance due to the high production needs, as the most of the agriculture land is placed in arid area [46]. Regarding the arable agriculture land, the total of 1.5 billion hectares were estimated, which about 250 million are under irrigation or about 17% total used land and for agricultural food production is estimated about 40% [47]. It was estimated that between 2000 and 2500 km^3 of water is consumed annually for irrigation. It is a well-known fact that on a global scale, about 70% of the affected quantities of water were consumed for agriculture [3], and irrigation is the main consumer of that water.

Water uptake relays on the osmotic potential of the soil solution, and the decrease in uptake can occur in the summer months on saline soils. At low temperatures, water uptake was reduced and then the plants experience physiological drought. This phenomenon can often occur in the spring; when due to the relatively high air temperature and low soil temperature, there is an imbalance in the water regime of plants, despite the fact that the soil contains sufficient amounts of water.

5.1 Soil water type

Water that results from precipitation or flooding was called surface water and it causes erosion and landslides. Free and bound water were most occurred states of soil water. The bounded water can occurred as: chemically bound water, hygroscopic water, membrane water, water in the form of water vapor and capillary water, while the free water includes gravitational, groundwater, and water in the form of ice [48]. Water in the soil is bound by various forces that the root system must overcome when adopted, so the water in the soil is divided into two classes: accessible and inaccessible.

Bound water in the soil is in the following forms:

1. Chemically bound water has no significance for the plant because it was bound within a solid lattice of minerals and as such belongs to the solid phase of the soil. Chemically bound water does not participate in physical processes and does not evaporate at a temperature of 100° C. It is present as constitutive and crystallized water. If water in the form of H^+ and OH^- ions enters the composition of different minerals, it was called crystal water. If it is bound to minerals as a molecule, then it is constitutional water. Chemically bound water is not available to the plant.

2. Hygroscopic water is water that is adsorbed on absolutely dry soil by surface forces at a relative humidity of less than 100%. The ability of soil particles to absorb relative moisture from the air was called hygroscopic water. Hygroscopic water in the soil is held at a high pressure of 50 bar; since the suction force of the root system is between 6 and 16 bar, it is inaccessible to plants. Maximum hygroscopic water is water constant but has no practical application other than being used to calculate other water constants. It is also adsorbed on dry soil by surface forces at a relative humidity of 95–100%.

3. Membrane or film water is located around soil particles and is used by plants only when the soil dries to membrane moisture. Membrane water binds to the surface of the particle after completion, that is, saturation of water binding to the maximum hygroscopy, if the particles can attract and to what extent there is available water. It was water bound by dipole forces that were weakening toward the periphery. Limestone water moves very slowly in the soil. According to Škorić [49], it is possible to divide film water into: stationary film water that corresponds to twice the value of hygroscopy and is inaccessible to plants, and mobile film water, where the water membrane is thick enough for water to move through plants is affordable. There are different zones: hygroscopic, which do forces greater than 50 bar, lentocapillary (6.25–50 bar) and membrane water of 0.50–6.25 bar hold, which is accessible to the plant.

4. Water in a gaseous state (water vapor) is physiologically useful if it turns into a liquid state by condensation, and it is a constant component of the soil air. The air in the soil saturated with water vapor with 98%. Water vapor in the soil moves from a warmer to a colder area or from an area of higher tension to lower tension.

5. Capillary water is very mobile and is of great importance in providing plants with water, as well as for physical and chemical processes in the soil. In dry climates, it is the only reserve for the plant, and the measures that allow the retention of capillary water were deep tillage, application of mulch, and cultivation. Capillary water is water that fills the narrowest pores of the soil due to the action of surface tension and occurs by increasing soil moisture. Capillary water is the most ecologically important form of water and is a basic factor in soil dynamics and fertility.

Free water in the soil is in the following forms: gravitational, groundwater, and ice-shaped water. For all forms of free water, one thing is common and that is when they are in liquid form, they move laterally and vertically under the influence of inclination or under the influence of gravity. The forces that hold water against soil particles are on the one hand the tension of moisture (surface, hydrostatic, and gravitational forces), and on the other hand, the osmotic pressure of the aqueous phase of the soil. Cohesion forces connect water molecules (hydrogen bridges and Van der Waals-London forces), whereas adhesion is responsible for their binding to soil particles and the formation of a double layer. Adhesion water is water that is located in the surface soil layer and retained by the forces of mutual molecular action between the soil particles and the absorbed water.

1. Gravitational or leach water is formed in occasion of fully saturated soil pores with water. In such a state of saturation, water seeps through macropores by gravity and is not bound to the soil. Gravitational water is the basic form of free water in the soil. After the precipitation lager, soil pores were fulfilled and, under the gravity, flows into the depth or sideways, down the slope. The size of the pores in the soil or soil texture was highly dependent on water flow movement in soil profile. If the soil has a lighter mechanical composition and contains a higher content of noncapillary pores, gravitational water will pass through the soil faster and less water will remain in the soil. There are two forms of gravitational water: fast gravitational water (for larger pores) and water that gradually drains away (for smaller pores). Gravitational water can be retained in the smallest noncapillary pores for several days during the wet period. She then swells with the help of her own weight.

2. Groundwater that is under a certain pressure in a water-permeable layer between water-impermeable layers. Groundwater is another form of free water and it formed if gravitational water encounters an impermeable layer of soil. Groundwater can also be associated with river water and then this water is close to the surface (1.0–2.5 m). The impact of groundwater on the soil depends on its depth and composition. Groundwater has no significance for plants if it is at great depth, but if it is located high and too close to the soil surfaces, it has a negative effect on plants. Groundwater is useful to plants when it is accessible to the roots due to capillary uplift or when the plants can absorb water from the deeper layer with long root system. Groundwater has a great impact on plants and soil if it contains soluble salts.

3. Ice form water or water in solid form is a special form of free water. This form of water is not of great importance for our climatic areas. Freezing and melting have a negative effect on the soil.

5.2 Movement of water in the soil

The three basic forms of water movement in liquid form are capillary movement, infiltration, and filtration. The movement of water is possible through unsaturated and saturated soil. Movement is possible descending, ascending, and lateral. The water direction of movement and speed were highly related to the occurrence state of the water, amount, texture, structure, porosity, the amount of organic matter, and the forces that cause the movement. The primary causes of soil water movement are capillary forces, gravity, and hydrostatic pressure.

The capillary movement of water occurs from the area of higher humidity to the area of lower humidity, that is, in unsaturated soil in micropores.

Infiltration is the uneven absorption by vertical and lateral motion into unsaturated soil, by the action of capillary, gravitational, and osmotic forces.

Filtration is the leaching of excess water from saturated soil into deeper layers through soil macropores, which causes gravity (and hydrostatic pressure).

Water in the soil moves in three basic directions: descending, ascending, and lateral. The descending flow of water is downward, with water draining freely through the macropores of the soil, primarily under the influence of gravitational force. This steady flow of water in cultural engineering corresponds to the concept of filtration. The ascending movement of water is ascending toward the soil surface and interpreted by capillary theory, membrane water theory, or potential difference (suction force—tension). According to capillary theory, water rises in profile due to the adhesion force that occurs between soil and water particles. Due to the adhesion, the capillaries fill and the water rises with the strength of the meniscus (adhesion) and due to the surface tension. According to the theory of membrane water, the ions in the outer diffuse shell have a suction power (osmosis) that fills the capillaries. Soil particles that have a thinner mantle accept water than those with a larger mantle. The last is the potential theory where water moves due to tension from a wetter to drier area. Lateral (lateral and radial) movement of water is interpreted by capillary theory, the theory of membrane water, and osmotic pressure, and the theory of potential.

The supply of soil with water from deeper soil layers depends on the type and composition but also on the method of soil cultivation depending on its purpose. Precipitation, which falls to the ground, is not equally abundant everywhere. Long-term hydrometeorological and agrometeorological measurements determine the daily, monthly, and annual average amount of precipitation for a particular area, which is logical to vary from place to place. Almost half of the water that enters the soil evaporates out into the atmosphere, about 2/6 is lost by leaching and runoff into depressions, streams, streams, and various standing waters, while only 1/6 is absorbed into the soil, which means that about 1/6 of the total precipitation and moisture that enters the soil remains available to the plants. Water retention is a very important factor in soil composition. This means that, for example, sandy soils differ significantly from clay soils in terms of moisture retention, absorption, and loss. Every soil has a certain degree of porosity because the soil made up of tiny fractions of sand and even tiny dust particles. Depending on the sand fraction, the pore size will also vary, or the permeability rate of such soil. Clay soil, on the other hand, consists largely of the finest particles of powder and clay, between which a very small number and diameter of pores were formed or have the ability to form, which limits the porosity of this type of soil.

5.3 Soil water constants

Water or hydrological constants defined as the equilibrium states between the suction force of soil particles and water. Most authors include the following in water constants:

- Hygroscopic water form
- Soil water capacities
- Maximum water capacity for soil
- Water retention and absolute water capacity for soil

- Field water capacity for soil
- Minimum water capacity for soil
- Humidity equivalent
- Lento-capillary point (humidity) or wilting point.

Water retention and water movement in soil were characterized by above mentioned water constants. Water constants were defined as the concept of water content, shape, and form in the soil related to the texture and structure of the soil, organic matter content, and applied agrotechnical measures. There were a number of water constants and various names were used in the literature and practice. Water constants that are of practical importance for the needs of soil hydro melioration and primarily for irrigation and drainage were listed here.

Hygroscopic water is, as mentioned earlier, the ability of soil particles to absorb relative moisture from the air. Humidity in contact with dry soil, allows the absorption of moisture that increases the volume of soil particles, until an equilibrium ratio is achieved. The established ratio represents the maximum absorption capacity with the achieved maximum hygroscopic effect, which for soils with sandy texture is 1%, for loamy soil texture up to 7%, and for clay soil texture up to 17% measured on dry matter. There is a difference between the maximum hygroscopy (Hm) according to Lebedev [50] and the hygroscopy according to Mitscherlich [51] (Hy). If the soil is placed in the conditions of complete saturation of the air with water vapor, then it will attract the maximum layer of hygroscopic water, this called maximum hygroscopy. Mitscherlich hygroscopy [51] corresponds to the moisture content, which is obtained by placing a soil sample in an evacuated desiccator above 10% sulfuric acid. The acid creates conditions of 96% relative humidity of the air that the soil absorbs. After establishing the equilibrium state by gravimetric method by weighing the moistened sample to hygroscopic moisture and completely dry soil, and by calculation in mass percentages, the moisture content corresponding to the hygroscopy is obtained. Hygroscopic moisture held in the soil, as mentioned earlier, by a suction force of 50 bar and is inaccessible to plants because the root of the plant has a suction force between 6 and 16 bar. The double value of Mitscherlich hygroscopy corresponds to the equilibrium state between the suction force of the plant root and the soil particles called the wilting point.

Soil water capacity presents a soil capacity for water retention in micropores after squeezing water from macropores under the influence of gravity. Depending on the method of determination, there were retention and absolute, field soil water capacity, minimum water capacity, and moisture equivalent according to Briggs and McLane [52].

Maximum soil water capacity (MWC) is a constant that represents the water content in the soil when water saturation is in maximum fulfilling the micropores and they are theoretically equal to the total porosity (**Figure 3**). When the maximum water capacity is then all other water constants were maximally realized. The state of maximum water capacity is short, especially in normal soils. Maximum water capacity is undesirable because anaerobic conditions occur, which puts the plants in a state of stress and puts maximum pressure on the metabolic activities of the plant. MWC occurs due to heavy rains and sudden melting of snow. In addition, the soil maximally saturated with water both when groundwater reaches the surface and during major floods. In rare situations and for a short period, the soil saturated to the MWC value. When the soil oversaturated with water, oxygen is lost in the soil and anaerobic conditions occur. This phenomenon is very harmful to soil and plants.

Excess water in the soil causes problems in the plant's oxygen supply (occurrence of anoxia = complete lack of oxygen and occurrence of hypoxia = reduced amounts of oxygen). Anoxia occurs more often if the temperature 12°C of the air is above 20°C, when the consumption of oxygen by breathing the roots of plants, soil fauna, and microorganisms is higher than at lower temperatures. Under such anaerobic conditions or at insufficient oxygen concentration, changes in the metabolism of plant tissue cells occur. Cell intoxication with alcoholic fermentation products and increase in cytoplasm acidity occurs. These phenomena can result in cell death. Plants suffering from the lack of oxygen show signs of wilting, due to the inability to active transfer water, and the leaves show epinastic growth (downward) due to increased ethylene synthesis. In such leaves, the concentration of abscisic acid increased, which initiates the closure of the shoot. This interrupts the transpiration flow and distribution of osmolytes and water to up ground plant parts which resulting in growth retardation.

The absolute soil water capacity according to Kopecky [53] corresponds to the water content in the micropores after complete saturation of the sample in a cylinder with a volume of 100 cm^3 and squeezing of excess water from the soil macropores after 24 hours. Horvat et al. [54] introduced the retention capacity of soil for water due to certain soil losses in determining the absolute capacity. According to this method, the soil sample in the cylinder, according to Kopecky [53] placed on a stand with filter papers whose edges immersed in water. The soil absorbs water and holds it in the micropores by adhesion, hydration, capillary, and surface tension forces. After squeezing, the excess water from the gravitational pores (the soil sample stands on the filter paper for half an hour) and the water content in volume % which corresponds to the retention capacity of the soil for water are is determined gravimetrically. The values of absolute and retention capacity of soil for water correspond to the soil layer above the groundwater level and are higher than those determined in field condition.

Field water capacity (FWC) is a condition where micropores are filled with water and macropores with air, after maximum saturation and seepage of free water under the influence of gravity. Soil moisture at FWC is retaining longer provided there is no evaporation or the influence of groundwater (capillary). For field water capacity, there are many names such as retention capacity, maximum water capacity, capillary capacity, and water retention at 0.33 bar. However, this constant (FWC) is determined in field conditions, and therefore, the name filed water capacity is the most appropriate (**Figure 6**). Field water capacity is extremely important because

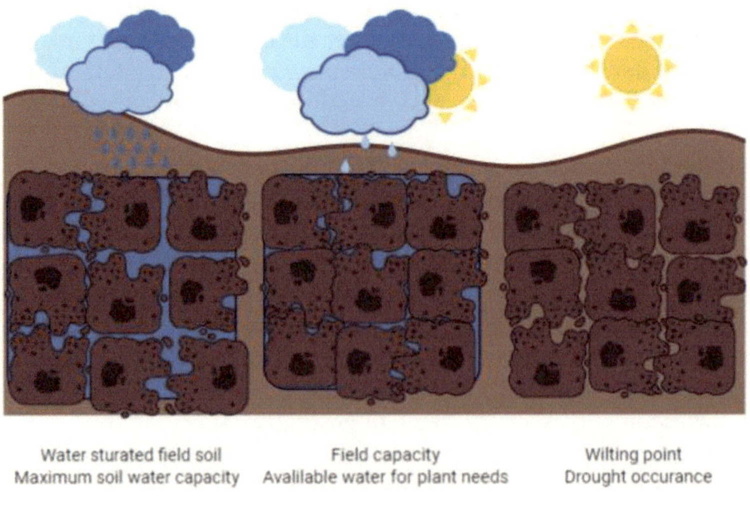

Figure 6.
Different water soil capacity.

its knowledge used in various calculations in the design and use of hydromelioration systems. FWC is especially important for irrigation because it is water constant without is impossible to accurate calculate the irrigation rate to moisten the active rhizosphere. FWC is also the largest amount of water that can be giving during irrigation because water above the FWC value is considered harmful to the plant. Irrigation practice depends on soil conditions, moisture and plant's requirements. Field capacity corresponds to the state of moisture that occurs in the soil immediately after rain, before evaporation and transpiration begin. It is expressed as the percentage of water in the soil that is found in such a state of humidity. Therefore, it corresponds to the water content in the well-drained, permeable soil from which all the gravitational water had drained. However, soil moisture that would correspond to the absence of gravitational water alone must be determined relatively soon after rain. This is possible only with light and well-drained soils, because in others, the process of rain infiltration through the ecological profile takes a long time, often several days. Meanwhile, after rain, the surface layers lose water by evaporation and transpiration, so the field capacity values that obtained after several days do not correspond to the actual water content of the soil.

Kramer [55] states that FWC will depend on soil temperature, that is, soil moisture in FWC will decrease as the soil temperature rises. Kirkham et al. [56] cite the influence of groundwater level, soil moisture depth, and impermeable layers in the soil on the FWC value. The author explains that the depth of soil moisture during infiltration will be greater the wetter the soil. Furthermore, the author states that the presence of an impermeable or less permeable soil layer increases the value of FWC.

The equivalent of moisture according to Briggs and McLane [52] is the water constant obtained by centrifuging a soil sample with a force of thousand times greater than the gravitational force. The value roughly corresponds to the field capacity of the soil for water. The lentocapillary point (Lkt) was defined as the lower limit of optimal soil moisture and corresponds to a soil water pressure of 6.25 bar (pF = 3.8). It is obtained by subjecting a saturated fine to the specified pressure in the pressure membrane. The wilting point (Tv) is the equilibrium state between the suction force of the root system and the soil particles and plants starts to lose turgor. A pressure that occurs in soil depends on available water and for this hydropedological constant is 15 bar (pF = 4.2). The point of wilting can be determined by vegetation experiments, using a pressure membrane and calculate from hygroscopy. Plant-accessible water is in the range between the value of soil water capacity and the wilting point (**Figure 7**)

Figure 7.
Philodendron in well turgid (left) and in less turgid condition (right).

and called physiologically active water. Within this interval, not all water is equally accessible, so the soil should maintain a moisture state between the water capacity and the lentocapillary point, which corresponds to the optimal moisture interval.

6. Drought stress

Drought is a common occurrence for many areas that recurs without noticeable regularity. Drought is one of the abiotic factors that have the greatest impact on agricultural production. It occurs when soil moisture decreases to an amount that negatively affects the yield and profitability of agricultural production [57]. Drought as an abiotic stress can directly affect agricultural production and even in some extreme situations lead to the complete destruction of yields. It is very important to distinguish the meaning of drought in agronomy and the definitions of drought in meteorology, hydrology, and the socio-economic concept of drought. The amount of physiologically active water in the soil, which is the only available water for plants, and the ratio of capillary and noncapillary pores in the soil are also important. The water content in the soil also depends on the texture of a particular soil type (i.e., on the water balance in the soil; fine sand has a lower possibility of water retention than clay loam). The ratio of humus in the soil is significant, because it has a great ability to absorb water (it acts like a sponge). The term drought should not be confused with the term "aridity," which refers to the permanent property of a naturally dry (waterless, arid) climate. A dry or dry area ("aryland") is an environment that was constantly, seasonally, or occasionally exposed to a significant lack of moisture. It was estimated that about 36% of the planet's continental surface, which is approximately 45 million km^2, can be classified as dry area. It is estimated that between 15 and 21% of the earth's population live in this area [4].

Regarding the extreme climatic conditions and rising temperatures, the need for irrigation of agricultural land is increasing; according to some statistic report, as much as 69% of drinking water is used in agriculture (irrigation of plants, watering livestock, etc.). FAO estimates that water consumption for plantation irrigation will increase by 5.5% from 2008 to 2050. These data show that the world's demand for water is increasing, and the increase in agricultural production results in higher consumption of drinking water. In order to reduce the need for irrigation, scientists studied the drought resistance of plants and sought to investigate the associated adaptation mechanisms. Geneticists are already creating varieties that contain drought-resistant genes; however, drought tolerance is only possible up to a certain percentage of moisture reduction. For example, chickpea (*Cicero arietinum* L.) is extremely drought resistant, its root is spindle-shaped and branched, and it reaches a depth of 1 m where there is a higher amount of moisture in the soil, and also many Mediterranean plant species are resistant to both high temperatures and lack of moisture in the soil.

Different time of drought periods also cause different changes in plants. Some plants increases synthesis of secondary metabolite as a response to drought occurrence resulting in leaf or fruit abscission and leading a plant cell, tissue, or organs to death (**Figure 8**). Transient wilting occurs during the hottest part of the day, when there is increased transpiration in plants. Permanent wilting occurs due to the low water content in the soil, root hairs die off, which blocks the plant's connection with nutrients from the soil (lost water can no longer be compensated). It is necessary to distinguish: desiccation (delay of drying ability to maintain hydrated tissue), tolerance to drying (retention of functions during drying), and avoidance of drought completion of the plant life cycle before drought occurs [13]. The point of wilting is the phase in which the plant begins to die, resulting in permanent death, and the plant when it reaches this stage (critical lack of moisture depends on the plant

Figure 8.
Citrus fruit abscission under the drought occurrence in critical phase (left) and leaf scrolling in tomato plant to avoid excess transpiration rate under stress condition (right).

species and type of cultivar) will not recover after irrigation. When a critical water deficit occurs, necrotic changes first appear on the plants, and finally, the lethal phase occurs. Plants have the greatest need for water during the growth phase, and during fruit formation, when plants are most sensitive to drought. According to the requirements of plants against water, they are divided into three main groups: hygrophytes (wetland plants), mesophytes (plants of temperate areas, which include most agricultural plants), and xerophytes (plants of arid areas).

One of the first defense mechanisms of the plant on drought is the reduction of the leaf surface, because with the reduction of water in the soil, the turgor pressure in the cells decreases, which ultimately results in a decrease in the concentration of cell content and cell volume. Turgor affects cell growth, and by reducing it, cell growth also decreases affecting the growth of the leaf surface and thus transpiration. Plants can reduce the surface area of transpiration by leaf scrolling and abscission (leaf rejection) or increased secretion of ethylene, which affects cell death. One of the most effective adaptations to drought is the closure of the stomata, to prevent further dehydration, this mechanism occurs when the plants have already fully developed their leaf surface. However, many plants lose water through the stomata because they remain open, and much of the water is lost through the epidermis by the cuticle, especially if the cuticle is thin.

Although it can be found in almost all parts of the world, the characteristics of drought vary from region to region. Defining drought is therefore difficult and depends on regional differences and needs, but also on the perspective from which this phenomenon is viewed. Regardless of the needs for which drought is defined, it is necessary that this definition includes the deviation of the current relationship between precipitation and evapo-transpiration in an area from the normal value of this relationship determined for a multi-year data set. It is also important to take into account the time distribution (precipitation regime, delay of the beginning of the rainy season, the relationship between precipitation, and phenological phases of the most important field crops in the observed area), as well as precipitation efficiency (precipitation intensity and number of rain episodes). Other climatic factors, such as high temperatures, high wind speeds and strengths, and low relative humidity, are often associated with drought in many parts of the world and can significantly worsen its consequences (**Figure 9**). Drought is an insidious natural disaster that, unlike other natural disasters, occurs slowly, lasts a long time, and affects large areas. It can be considered from four aspects (meteorological, hydrological, agricultural, and socio-economic).

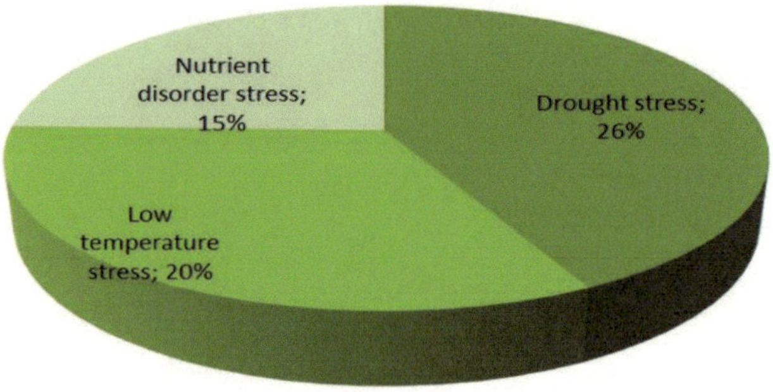

Figure 9.
Current trends of most abundant abiotic stress occurrence in agriculture sector estimated for around 50% of world's arable land.

Short-term water shortage over a period of several weeks in the surface layer of the soil, which occurs at a critical time for plant development, can cause agronomic drought. The agronomic droughts may lag behind meteorological droughts, depending on the condition of the surface layer soil. High temperatures, low relative humidity, and wind amplify the negative consequences agronomic droughts. The agronomic droughts, precipitation deficits are taken into account along with the physical and biological aspects of plants, interactions within the soil-plant-atmosphere system and the balance between plants' water needs and available water reserves, which may result in declining yields. Beside agronomical drought, we can also distinct the meteorological and hydrological drought.

Meteorological drought occurs as a consequence of lack or complete absence of precipitation over a long period of time in a certain area. This deficiency is defined as the deviation of precipitation from normal, that is, from the multi-year average. Meteorological drought can develop abruptly and stop abruptly.

Deficit of precipitation over a long period of time affects surface and groundwater supplies: to the flow of water in rivers and streams, level of water in lakes, and level of groundwater. When flows and levels decrease, we talk about hydrological drought. The onset of hydrological drought may lag a few months behind the beginning of the meteorological drought, but also continue after the end of the meteorological droughts. Finally, socio-economic drought could be defined as an event when the need for water is greater than the possibility to provide it with agrotechnical measures. The mentioned concept of drought reflects a strong connection between drought and human activities. The droughts in recent years have had a significant impact on the economies and environment of agricultural production, increasing the vulnerability of society as well as a whole ecosystem.

6.1 Mechanisms of plant adaptation to drought

The lack of water especially in agricultural sector presents important limitation of world food production. The strategies for promoting the mechanisms of plant tolerance to drought:

a. Desiccation delay is the ability to maintain hydration of the tissue

b. Desiccation tolerance is the retention of cell functions during drought

c. Drought avoidance is the end of the plant vegetation cycle before drought occurs.

It is already clear from this division that the mechanisms and strategies of drought resistance may differ. Water deficiency could be explained as amount of water in a cell tissue that is lacking until optimal hydration. When water deficit gradually occurs, the impact of water scarcity on plant growth and development comes to the fore. The basic acclimatization strategies that occur in conditions of water scarcity are:

- reduction of leaf area,
- leaf rejection (abscissa),
- increased root growth,
- retaining the stomata,
- osmotic adjustment,
- thickening of the cuticle.

Decrement of water plant status affects the turgor pressure, which also decreases. As turgor falls, cell volume decreases, cell contents become more concentrated, and cell membrane becomes less tense and thicker. Cell growth if highly influenced by turgor and consequently declining of turgor will restrict cell growth. In addition to reducing turgor, the lack of water also reduces the elasticity of cell walls, which also affects cell growth. Decreased cell growth results in smaller leaves, that is, reduced leaf area. Reduced leaf area helps conserve water because smaller leaves breathe less (lose water more slowly). Therefore, a plant usually starts to decrease leaves surface and afterwards its number as a result on water stress conditions. The leaves age and fall off faster (there is an increased synthesis of ethylene, which encourages leafs fall).

Beside leaves, plant root is also susceptible to the lack of water. The balance between the uptake of water through the root and the photosynthetic activity of the aboveground part influences the ratio of the root mass and aboveground part. Simply, the aboveground organs will grow as long as the root supplies them with sufficient water and nutrients, and conversely, the root will grow as long as the aboveground organs supply it with sufficient assimilates. As already mentioned, water deficiency cause leaf area and number of leaves decrease in the early phase, while the intensity of photosynthesis tends to remain unchanged. Reducing the leaf area allows less water consumption, but also energy, so more carbohydrates were translocated to the root and allowed it to grow. However, in dry soil, the root tip loses turgor very quickly, so the root grows where the soil is still moist. As water scarcity (drought) progresses, stress most often occurs. Stress caused by the lack of water due to drying out of the upper soil layers so the plants develop deeper roots. The development of deeper roots is the also one of the reaction pathways to the drought. Increased root elongation during drought requires translocation of assimilates from aboveground organs to the root. In the generative phase, a significant outflow is represented by fruits (assimilates are spent on fruit growth), so the roots get less assimilates. Therefore, if the stress of water deficiency occurs in the generative phase of plant development, the effect of enhanced root growth will be less pronounced [10].

In conditions of intense stress (rapid) occurrence as a result of water deficit or the phase where plants have already developed the maximum leaf area, other mechanisms activated to prevent the plant from drying out. One of the most important such mechanisms is the closure of the stoma, which reduces transpiration, that is, water loss. Therefore, the “third line” of drought protection can be considering the mechanism of stomata closuring. The change of turgor in the guardian cells regulates the opening and closing mechanism of the leaves stomata. The guardian cells are modified cells of the leaf epidermis and that is why they can lose turgor as they loss of water (transpiration) into the atmosphere. Such a way of holding, stomata (due to direct water loss and falling turgor) called hydro-passive stoma detention. The second mechanism called hydro-active stomata closuring occurs in conditions when the whole leaf and/or root is dried. This mechanism is triggered by metabolic processes in the stomata cells. Concentration decrement of osmotic active substances in the stomata results in water release in guardian cells and turgor pressure decrease. Hydro-active detention of the stoma occurs due to a decrease in osmotic potential, which leads to the release of water and a decrease in turgor in the guardian cells. The abscisic acid (ABA) affects the decrement of concentration in osmotic active substance in the stoma. Abscisic acid in very low concentrations is constantly synthesized in meso-philic cells and accumulates in chloroplasts. Under conditions of mild dehydration of mesophylls, two processes activated: (1) part of the abscise acid stored in chloroplasts is released into the apoplast (intercellular spaces) of mesophilic cells and then the transpiration current of water carries ABA to the guardian cells and (2) the synthesis of ABA intensifies and its higher concentrations accumulate in the apoplast of the leaf. This second process (ABA synthesis) prolongs, that is, maintains the process of coupling retention that occurs due to the release of ABA from the chloroplast. In addition, during stress caused by the lack of water, chemical signals (ABA) transmitted from the root to the leaf, leading to the stomata closing. In fact, the conductivity of the shoot indirectly was managed by the soil water status rather than the water status of the leaves, afterward root show high sensitivity respond to lac of soil moisture.

As the amount of water in the soil decreases, the water potential of the soil decreases. In condition when root water potential is more negative then soil water potential plants can receive necessary water. The process of accumulation of solutes (osmotic active substances) in cells presents osmotic adaptation. Thus, the cell reduces the water potential (the water potential becomes more negative) without a significant change in the turgor or volume of the cell. During osmotic adaptation, various solutes accumulate in the cell, mainly sugars, organic acids, amino acids, and inorganic ions (especially K^+). Ions mainly accumulate in the vacuole because their high concentration in the cytosol can inhibit many enzymes. However, due to the increased concentration of ions in the vacuole, and in the cytoplasm, there must be an increase in the concentration of solutes in order to maintain the balance of water potential within the cell. Dissolved substances that accumulate in the cytoplasm called compatible osmotic active substances and are substances that do not inhibit enzymes. Compatible osmotic active substances were proline as amino acids, and sugar alcohols and glycine betaine as amine. The synthesis of compatible osmotic substances also occurs under conditions caused by increased salinity. The osmotic adjustment takes place slowly over several days. Osmotic adjustment of leaves can provide turgor maintenance with lower water potential compared to leaves that have not undergone osmotic adjustment. The maintenance of turgor enables the normal course of cell growth (cell elongation) and greater stomata conductivity at lower water potential, which leads to the conclusion that the osmotic adaptation is actually a process of acclimatization.

The cuticle is a waxy coat located above the epidermal cells, and its function is to reduce water loss (cuticular transpiration). In conditions of lack of water, plants often synthesize a thicker cuticle. The thicker cuticle also reduces the entry of CO_2, but photosynthesis usually remains unchanged because the epidermal cells located below the cuticle do not conduct photosynthesis (CO_2 for photosynthesis enters the leaf through the stoma). Cuticular transpiration makes up only 5–10% of the total transpiration, so the thickness of the cuticle is important only in conditions of more severe drought or when the cuticle is damaged (e.g., due to wind). The lack of water reduces the intensity of photosynthesis although this process does not react as pronouncedly to lack of water, as is the case with leaf area. The reason for this is that photosynthesis is not as sensitive to turgor decline as cell growth. However, due to the detention of the stoma under drought stress conditions, the entry of CO_2 into the leaf reduced and the intensity of photosynthesis decreases.

Each plant species has its own root growth characteristics, which can be substantially, modify by the plant's environment. In most arable land used in agriculture, the roots occupy the largest volume of soil, although it is dictated by the physical and chemical properties of the soil. The main roots and root hairs are responsible for transporting water through the conductive elements, while the lateral roots play an important role in the absorption of water and minerals [58–60]. The roots are under the extremely high influence of soil factors, its buffering abilities, surrounding environment as one of the most important parameters that reflected in the development of the roots [58]. The soil water and nutrient content, soil type, vegetation, and agrotechnical measures strongly reflect root growth [61–63, 60, 64]. In the early growth phase, roots behave as sink for assimilates which are distributed in vegetative parts of plant until generative phase occurs this role become weaker [61]. Richards [65] found that wheat root shows significant changes if grown in water deficit conditions and also this state reflects root development. Dry soil my intensify growth of root in depth searching for water [66] but elongation of root cells could be restricted [67].

The balance between the aboveground and underground part of the plant is very important in terms of regulating the status of water in the tissue, and it was found that some plants tolerate drought conditions better if the ratio between aboveground and underground plant parts were lower [68]. Under favorable water conditions in the soil and the wheat phase of vegetation, better rooting and increase in root volume can occur [69]. Taylor and Klepper [70] found that minor changes in water potential did not significantly affect the increase in plant root volume.

A large number of studies regarding the influence of water stress on the plant conclude that not only the root results in certain changes but also certain changes occur in the aboveground part of the plant. Beside arid area also water saturated soil can provide certain changes on root (**Figure 10**) and affect root development as

Figure 10.
Salix sp. *root adaptation to water level changing or species tolerant to hypoxic conditions by forming abounded fibrous root.*

a result of anaerobic conditions [71] even some species have adapted to this condition developing a special root [72]. The reaction of the plant to stressful conditions in most cases was reflected in the unfavorable status of water in plant tissues and organs. The main mechanism of plant cell resistance to stress is to maintain favorable turgor pressure in cells, tissues, and organs (stress is first reflected in important physiological processes in the plant such as photosynthesis, respiration, transpiration, nutrient, and water uptake) [64, 73, 74].

The roots of plants are most often adapted to stressful conditions in a way that changes the structure of the tissue and changes in root volume occur, which was closely related to the reclamation of the osmotic activity of root cells. Cells most often try to protect cells from protoplast dehydration through metabolic exchange of water molecules and some osmotic active substance that result tissue adaptation to newly state that occurred (wall thickness, cell size, osmolite concentration, etc.) [66, 75, 76].

7. Conclusion

Life on earth originated in water and depends on water which is necessary for the life of all beings. Plants are constantly receiving and excreting water that has a number of biochemical and physiological functions in the plant organism. Plant species as well as individual plant parts differ in terms of water demand. Of all the factors that affect the growth and development of a plant, water have become the most important and common limiting factor. Plant tolerance to abiotic stresses, especially drought stress is a very complex process consisting of a series of physiological, biochemical, and genetic adaptations, that is, evolutionary adaptations of certain plant species. The biggest challenge is to determine exactly which genes activate certain chemical compounds and substances which control the plant's response on drought stresses and how exactly the plant physiologically manages to respond to stimuli. The agricultural production is dependence on weather conditions as of today it is becoming more and more pronounced due to the increasingly frequent weather extremes caused by the climate changes. The lack of precipitation or their uneven distribution, pronounced dry periods almost regularly monitored by above-average temperatures, and extreme weather conditions such as hail, floods, and strong winds negatively affect agricultural production. Yield drop or total loss of yield and financial losses are the main direct consequences of adverse weather conditions. One of the most important factors that affect plant habitat and available source of water is soil capacity and physical properties of soil. Some agrotechnical measures can improve soil water capacity but in fact for all agricultural produces always remain as challenge—how to manage water in a way to provide optimal crop yields during the vegetation season.

Acknowledgements

The author wishes to express gratitude to the Ministry of Foreign Affairs of the Czech Republic (Project No. 7/2020/01 "Increasing scientific research capacities and support of education at the University of Mostar and University of Dzemal Bjedic Mostar").

Conflict of interest

The authors declare no conflict of interest.

Author details

Adrijana Filipović
Faculty of Agriculture and Food Technology, University of Mostar, Mostar, Bosnia and Herzegovina

*Address all correspondence to: adrijan.filipovic@aptf.sum.ba

References

[1] Häder D-P et al. Effects of UV radiation on aquatic ecosystems and interactions with other environmental factors. Photochemical & Photobiological Sciences. 2015;**14**(1):108-126. DOI: 10.1039/C4PP90035A

[2] Pimentel D et al. Water resources: Agricultural and environmental issues. Bioscience. 2004;**54**(10):909-918. DOI: 10.1641/0006-3568(2004)054[0909:WRAAEI]2.0.CO;2

[3] Lozán JL, Meyer S, Karbe L. Water as the basis of life. In: Lozan JL, Grassl H, Hupfer P, Menzel L, Schonwiese C, editors. Global Change: Enough Water for All? 2nd ed. Hamburg: Wissenschaftliche Auswertungen/GEO; 2007:19-25. Available from: https://www.academia.edu/26410592/1_Water_as_the_basis_of_life

[4] Bonacci O. Suše-nekoć i danas. Hrvatske vode; 2015. pp. 133-141, 23, 92. Available from: https://www.voda.hr/sites/default/files/pdf_clanka/hv_92_2015_133_bonacci.pdf

[5] Tsindos S. What drove us to drink 2 litres of water a day? Australian and New Zealand Journal of Public Health. 2012;**36**(3):205-207. DOI: 10.1111/j.1753-6405.2012.00866.x

[6] Brini E, Fennell CJ, Fernandez-Serra M, Hribar-Lee B, Lukšič M, Dill KA. How water's properties are encoded in its molecular structure and energies. Chemical Reviews. 2017;**117**(19):12385-12414. DOI: 10.1021/acs.chemrev.7b00259

[7] Engin K. Facing the Challenges: Case Studies and Indicators: UNESCO's Contributions to the United Nations World Water Development Report 2015. UNESCO Publishing; 2015. Available from: http://www.unesco.org/new/fileadmin/MULTIMEDIA/HQ/SC/images/__WWDR_2015_CaseStudies_Indicators_web_02.pdf

[8] Mishra RK, Dubey SC. Fresh water availability and it's global challenge. International Journal of Engineering Science Invention Research & Development. 2015;**II**(VI):351-407. Available from: www.ijesird.com. e-ISSN: 2349-6185

[9] Aung K, Jiang Y, He SY. The role of water in plant-microbe interactions. The Plant Journal. 2018;**93**(4):771-780. DOI: 10.1111/tpj.13795

[10] Pevalek-Kozlina B. Fiziologija bilja. Zagreb: Profil International; 2003

[11] Taiz L, Zeiger E, Møller IM, Murphy A. Plant Physiology and Development. 6th ed. Sinauer Associates, Oxford University Press; 2018. Available from: http://6e.plantphys.net/

[12] Dubravec KD, Regula I. Fiziologija bilja, Školska knjiga Zagreb; 1995. Available from: https://www.knjiga.ba/fiziologija-bilja-b1150.html

[13] Lazarević B, Poljak M. Fiziologija bilja. Sveučilište u Zagrebu. Zagreb: Agronomski fakultet; 2019

[14] Devlin RM. Plant Physiology. Van Nostrand Reinhold Company; 1975. ISBN 10:0442220936, ISBN 13:9780442220938

[15] Mader SS. Biology: Plant Structure and Function, 4th Revised ed. Dubuque, Iowa: William C Brown Pub; 1993

[16] McCully ME. Roots in soil: Unearthing the complexities of roots and their Rhizospheres. Annual Review of Plant Physiology and Plant Molecular Biology. 1999;**50**(1):695-718. DOI: 10.1146/annurev.arplant.50.1.695

[17] Sutcliffe JF. Plants and Water. Edward Arnold; 1979. Available from: https://www.cabdirect.org/cabdirect/abstract/19680604052

[18] Locher JT, Brouwer R. Influence of Different Root Temperature on Transpiration and Exudation of Young Maize Plants. Wageningen: Jaarb. I.B.S. 1965. pp. 280-295, 57-65

[19] Mengel K, Kirkby EA. Principles of Plant Nutrition. 5th ed. Springer Netherlands; 2001. Available from: https://link.springer.com/book/10.1007/978-94-010-1009-2

[20] Slatyer RO, Slatyer RO. Plant-Water Relationships. Academic Press; 1967. DOI: 10.1126/science.158.3805.1171-a

[21] van den Honert TH. Water transport in plants as a catenary process. Discussions of the Faraday Society. 1948;**3**:146-153. DOI: 10.1039/DF9480300146

[22] Kaiser WM. Effects of water deficit on photosynthetic capacity. Physiologia Plantarum. 1987;**71**(1):142-149. DOI: 10.1111/j.1399-3054.1987.tb04631.x

[23] Giménez C, Gallardo M, Thompson RB. Plant–Water Relations. In: Reference Module in Earth Systems and Environmental Sciences. Elsevier; 2013. Available from: https://plantstomata.wordpress.com/tag/r-b-thompson/

[24] Turner NC et al. Osmotic adjustment in chickpea (*Cicer arietinum* L.) results in no yield benefit under terminal drought. Journal of Experimental Botany. 2007;**58**(2):187-194. DOI: 10.1093/jxb/erl192

[25] Chen H, Jiang J-G. Osmotic adjustment and plant adaptation to environmental changes related to drought and salinity. Environmental Reviews. 2010;**18**:309-319. DOI: 10.1139/A10-014

[26] Johansson I, Karlsson M, Johanson U, Larsson C, Kjellbom P. The role of aquaporins in cellular and whole plant water balance. Biochimica et Biophysica Acta (BBA)–Biomembranes. 2000;**1465**(1-2):324-342. DOI: 10.1016/S0005-2736(00)00147-4

[27] Gomes D, Agasse A, Thiébaud P, Delrot S, Gerós H, Chaumont F. Aquaporins are multifunctional water and solute transporters highly divergent in living organisms. Biochimica et Biophysica Acta. 2009;**1788**(6):1213-1228. DOI: 10.1016/j.bbamem.2009.03.009

[28] Jeje AYA, Zimmermann MH. Resistance to water flow in xylem vessels. Journal of Experimental Botany. 1979;**30**(4):817-827. DOI: 10.1093/jxb/30.4.817

[29] Kramer PJ, Boyer JS. Water Relations of Plants and Soils. Academic Press; 1995. Available from: http://udspace.udel.edu/handle/19716/2830

[30] 361714.pdf [Online]. Available from: https://edepot.wur.nl/361714#page=52 [Accessed: 05 June 2020]

[31] Bollard EG. Transport in the xylem. Annual Review of Plant Physiology. 1960;**11**(1):141-166. DOI: 10.1146/annurev.pp.11.060160.001041

[32] Sack L, Holbrook NM. Leaf hydraulics. Annual Review of Plant Biology. 2006;**57**(1):361-381. DOI: 10.1146/annurev.arplant.56.032604.144141

[33] Sack L, Tyree MT. Chapter 5–Leaf hydraulics and its implications in plant structure and function. In: Holbrook NM, Zwieniecki MA, editors. Vascular Transport in Plants. Burlington: Academic Press; 2005. pp. 93-114

[34] Kim HK, Park J, Hwang I. Investigating water transport through the xylem network in vascular plants.

Journal of Experimental Botany. 2014;**65**(7):1895-1904. DOI: 10.1093/jxb/eru075

[35] Jezek M, Blatt MR. The membrane transport system of the guard cell and its integration for stomatal dynamics. Plant Physiology. 2017;**174**(2):487-519. DOI: 10.1104/pp.16.01949

[36] Lawson T, Blatt MR. Stomatal size, speed, and responsiveness impact on photosynthesis and water use efficiency. Plant Physiology. 2014;**164**(4):1556-1570

[37] Vialet-Chabrand SRM, Matthews JSA, McAusland L, Blatt MR, Griffiths H, Lawson T. Temporal dynamics of stomatal behavior: Modeling and implications for photosynthesis and water use. Plant Physiology. 2017;**174**(2):603-613. DOI: 10.1104/pp.17.00125

[38] von Möhl H. Welche Ursachen bewirken die Erweiterung und Verengung der Spaltöffnungen? Botanische Zeitung. 1856;**14**(697-704):713-721

[39] Tyree MT, Ewers FW. The hydraulic architecture of trees and other woody plants. The New Phytologist. 1991;**119**(3):345-360. DOI: 10.1111/j.1469-8137.1991.tb00035.x

[40] Tyree M. The Cohesion-Tension theory of sap ascent: Current controversies. Journal of Experimental Botany. 1997;**48**:1753-1765. DOI: 10.1093/jexbot/48.315.1753

[41] Ogburn RM, Edwards EJ. The ecological water-use strategies of succulent plants. In: Advances in Botanical Research. Vo.l 55. Elsevier; 2010. pp. 179-225. DOI: 10.1016/S0065-2296(10)55004-3

[42] Noy-Meir I. Desert ecosystems: Environment and producers. Annual Review of Ecology and Systematics. 1973;**4**(1):25-51. DOI: 10.1146/annurev.es.04.110173.000325

[43] Reynolds JF, Kemp PR, Ogle K, Fernández RJ. Modifying the 'pulse-reserve' paradigm for deserts of North America: Precipitation pulses, soil water, and plant responses. Oecologia. 2004;**141**(2):194-210. DOI: 10.1007/s00442-004-1524-4

[44] Chen X, Hu Q. Groundwater influences on soil moisture and surface evaporation. Journal of Hydrology. 2004;**297**:285-300. DOI: 10.1016/j.jhydrol.2004.04.019

[45] Beltrão J, Antunes Da Silva A, Asher JB. Modeling the effect of capillary water rise in corn yield in Portugal. Irrigation and Drainage Systems. 1996;**10**(2):179-189. DOI: 10.1007/BF01103700

[46] Simunic D, Ye M, Zhang P. Improving audit value with country specific interpretations after International Auditing Standards are adopted. SSRN Electronic Journal. 2014. DOI: 10.2139/ssrn.2402447

[47] van Hofwegen P, Svendsen M. A Vision of Water for Food and Rural Development: Final [Online]. 2000. Available from: https://agris.fao.org/agris-search/search.do?recordID=NL2000004754 [Accessed: 07 June 2020]

[48] Vučić NV, Putanov P. Vojvođanska akademija nauka i umetnosti, Vodni, vazdušni i toplotni režim zemljišta. Novi Sad: Vojvođanska akademija nauka i umetnosti; 1987

[49] Škorić A. Tipovi naših tala. Zagreb: Liber; 1977

[50] Lebedev AF. Pochvennye i gruntovye vody. Izd-vo Akademii nauk SSSR; 1936. Available from: http://scholar.google.com/scholar_lookup?&title=Pochvennye%20i%20gruntovye%20vody&publication_year=1936&author=Lebedev%2CA.F.

[51] Mitscherlich EA. Bodenkunde für Land- und Forstwirte. Berlin: P. Parey; 1905

[52] Briggs LJ, McLane JW. The Moisture Equivalents of Soils. U.S. Government Printing Office; 1907. Available from: https://ufdc.ufl.edu/AA00025979/00001

[53] Kopecký J. Die physikalischen Eigenschaften des Bodens: Wasserkapazität, Porosität, Spezifisches Gewicht, Luftkapazität und Durchlässigkeit, ihre Bedeutung und Bestimmung. Prag: Buchdruckerei der "Politik"; 1904

[54] Horvat I, Tomažič G, Horvatić S, Em H, Gračanin M, Maksić B. Priručnik za zipološko istraživanje i kartiranje vegetacije. Zagreb: Nakladni Zavod Hrvatske; 1950

[55] Kramer PJ. Water Relations of Plants. Academic Press; 1983. Available from: https://www.sciencedirect.com/science/book/9780124250406

[56] Kirkham S, Lam S, Nester C, Hashmi F. The effect of hydration on the risk of friction blister formation on the heel of the foot. Skin Research and Technology. 2014;**20**(2):246-253. DOI: 10.1111/srt.12136

[57] Mannocchi F, Todisco F, Vergni L. Agricultural drought: Indices, definition and analysis. In: Rodda JC, Ubertini L, editors. The Basis of Civilization – Water Science? Proceedings of the UNESCO/IAHS/IWIIA Symposium. Rome, Italy: IAHS Publ. 2004;**286**:246-254

[58] O'Toole JC, Bland WL. Genotypic variation in crop plant root systems. In: Advances in Agronomy. Vol. 41. N. C. Brady: Ed. Academic Press; 1987. pp. 91-145

[59] Tardieu F, Katerji N. Plant response to the soil water reserve: Consequences of the root system environment. Irrigation Science. 1991;**12**(3):145-152. DOI: 10.1007/BF00192286

[60] Yamaguchi T, Moldrup P, Rolston DE, Ito S, Teranishi S. Nitrification in porous media during rapid, unsaturated water flow. Water Research. 1996;**30**(3):531-540. DOI: 10.1016/0043-1354(95)00206-5

[61] Hamblin AP, Tennant D. Root length density and water uptake in cereals and grain legumes: How well are they correlated. Australian Journal of Agricultural Research. 1987;**38**(3): 513-527. DOI: 10.1071/ar9870513

[62] Hamblin A, Tennant D, Perry MW. The cost of stress: Dry matter partitioning changes with seasonal supply of water and nitrogen to dryland wheat. Plant and Soil. 1990;**122**(1): 47-58. DOI: 10.1007/BF02851909

[63] Lipiec J, H→ing;kansson I, Tarkiewicz S, Kossowski J. Soil physical properties and growth of spring barley as related to the degree of compactness of two soils. Soil and Tillage Research. 1991;**19**(2):307-317. DOI: 10.1016/0167-1987(91)90098-I

[64] Royo A, És RAU, Playán E, Ortiz R. Salinity–grain yield response functions of barley cultivars assessed with a drip-injection irrigation system. Soil Science Society of America Journal. 2000;**64**(1):359-365. DOI: 10.2136/sssaj2000.641359x

[65] Richards RA. Crop improvement for temperate Australia: Future opportunities. Field Crops Research. 1991;**26**(2):141-169. DOI: 10.1016/0378-4290(91)90033-R

[66] Klepper B, Taylor HM, Huck MG, Fiscus EL. Water relations and growth of cotton in drying soil1. Agronomy Journal. 1973;**65**(2):307-310. DOI: 10.2134/agronj1973.00021962006500020036x

[67] Taylor HM. Managing root systems for efficient water use: An overview. In: Limitations to Efficient Water Use

in Crop Production. John Wiley & Sons, Ltd; 2015. pp. 87-113. Available from: https://acsess.onlinelibrary.wiley.com/doi/pdf/10.2134/1983.limitationstoefficientwateruse.c6

[68] Grzesiak S, Grzesiak MT, Filek W, Hura T, Stabryła J. The impact of different soil moisture and soil compaction on the growth of triticale root system. Acta Physiologiae Plantarum. 2002;**24**(3):331-342. DOI: 10.1007/s11738-002-0059-8

[69] Sharma BR, Chaudhary TN. Wheat root growth, grain yield and water uptake as influenced by soil water regime and depth of nitrogen placement in a loamy sand soil. Agricultural Water Management. 1983;**6**(4):365-373. DOI: 10.1016/0378-3774(83)90055-0

[70] Taylor HM, Klepper B. Water relations of cotton. I. Root growth and water use as related to top growth and soil water content 1. Agronomy Journal. 1974;**66**(4):584-588. DOI: 10.2134/agronj1974.00021962006600040031x

[71] Cheng W, Coleman DC, Box JE. Measuring root turnover using the minirhizotron technique. Agriculture, Ecosystems and Environment. 1991;**34**(1):261-267. DOI: 10.1016/0167-8809(91)90113-C

[72] Davis TD, Haissig BE. Biology of Adventitious Root Formation. Springer Science & Business Media; 1994. DOI: 10.1007/978-1-4757-9492-2

[73] Linkemer G, Board JE, Musgrave ME. Waterlogging effects on growth and yield components in late-planted soybean. Crop Science. 1998;**38**(6). DOI: 10.2135/cropsci1998.0011183X003800060028x

[74] Przywara G, Stępniewski W. The influence of waterlogging at different temperatures on penetration depth and porosity of roots and on stomatal diffusive resistance of pea and maize seedlings. Acta Physiologiae Plantarum. Dec. 1999;**21**(4):405-411. DOI: 10.1007/s11738-999-0013-0

[75] Levitt J. Responses of plants to environmental stress. In: Chilling, Freezing, and High Temperature Stresses. New York, London, Toronto, Sydney, San Francisco: Academic Press; 1980:**1**. Available from: https://trove.nla.gov.au/work/10246393

[76] Poljakoff-Mayber A, Bar-Nun N, Hasson E, Heichal O. Respiratory carbohydrate metabolism of different pea varieties under saline conditions. Botanical Gazette. 1981;**142**(4):431-437. DOI: 10.1086/337243

Chapter 5

Soil Salinity and Its Management

Muthuraman Yuvaraj, Kasiviswanathan Subash Chandra Bose, Prabakaran Elavarasi and Eman Tawfik

Abstract

Soil salinity is a growing threat all over the world due to its toxic effect to reduce soil fertility and water uptake in the crops. An average of 418 million ha soil is saline in nature. Various climatic, geomorphic and rainfall pattern causes which involved in saline soil formation. To reduce the toxic effect proper management of saline soil is required. Irrigation water also a major concern regarding soil salinity management. Saline irrigation water enhances and maintains the severity soil salinity. Crop production aspects root zone salinity provides a strong negative impact on soil fertility. Salinity causes the reduction in nutrient ion, and water uptake has a significant negative effect on crop yields. Soil and water salinity interactions and their influence on crop growth and management of salinity are deliberated in this chapter.

Keywords: salinity, management, soil fertility, crop growth

1. Introduction

All over the world more than hundred countries approximately 418 million hectare of saline soils are present. Asia alone contribute 46% of soils are salt affected the world. Arid and semi-arid regions of the word highly affected by salinisation 6.27 percentage of soil in Asia affected by salinity. In India 10 mha salt affected soil or 5.5 mha saline and 3.8 mha are sodic soils [1, 2]. Saline soil mainly consist soluble salts like chloride and sulphates of calcium (Ca), magnesium (Mg), sodium (Na), potassium (K), bicarbonate (HCO_3^-), carbonate (CO_3^{2-}) and nitrate (NO_3^-) also present. The EC (electrical conductivity) of saline soil is less than 4 dS m^{-1} (deci Siemens/meter), ESP (exchangeable sodium percentage) is less than 15% and pH is less than 8. Less soluble salts like calcium carbonate and calcium sulphate also present. Saline soil having more ionic salt species like Ca^{2+} which flocculates the soil when its dominated by Na which disperse the soil. Saline soils are mostly Ca^{2+} dominated flocculated and well aerated in nature. Dispersion and flocculation also based on clay content of the soil.

2. Process of soil salinisation

There are two forms of salinisation process are present. They are primary or natural salinisation and secondary or anthropogenic salinisation. Due to various hydropedological, geomorphic and climatic factors involved in primary salinisation. Hydeopedological factors like weathering of basic rocks like basalt, gabbro and dolerite, etc. Hydrological factors like high annual moisture flux, climatic factors like low precipitation and high temperature cause more evapotranspiration and geomorphic factor like low relief are the cause of primary salinisation [3–5]. Secondary salinisation due to shallow water level, poor quality irrigation water, over irrigation with improper drainage make increase in water table and discharge of salt near soil surface, over use of ground water in coastal regions sea water intrusion takes place. Finally untreated industrial effluents and waste water having high dissolved salts are the reasons for secondary salinisation [6–9].

3. Measure of analysing soil salinity

Electrical conductivity (EC) is a measure of soil salinity and it's measured by EC meter. EC is reciprocal to the resistance of the conductor. Resistance is directly proportional to the distance between the electrode and inversely proportional to the cross sectional area of the conductor. Resistance expressed as ohms cm^{-1} conductivity expressed m mhos cm^{-1}. Conductivity depends on concentration of the ions in the solution, temperature, valence and length between two electrodes. The conductivity is directly proportional to the, concentration of the solution, solution temperature and valance and inversely proportional to length between two electrodes [10].

Resistance is calculated by ohms law.

$$R = V / I.$$

$$G = 1 / R.$$

where V is the voltage; I, current (ampere); R, resistance of the solution; G, conductance.

Conductance – Reciprocal of electrical resistance of a solution between two electrodes [11].

3.1 Saturation extract (ECe)

In this method involves saturation and subsequent extraction of soil water under partial pressure in order to quantify the salts in the extracted liquid phase. Crop growth response mainly depends on ECe (saturation paste) conversion of EC to ECe on different structural soil important in plant growth point of view. Sand particles can able to hold or adsorb more ions which leads to concentration of salinity will be more in the sandy than clay soil when the amount of salt ion will be the same [12].

3.2 Ranges of soil salinity

Non saline	Slightly saline	Moderately saline	Strongly saline	Extremely saline
0–2	2–4	4–8	8–16	>16

Electrical conductivity (EC) in dS m^{-1} [13].

4. Irrigation water salinity

Irrigation water salinity also a major concern regarding secondary salinisation and crop production (**Table 1**). Saline water highly affects germination,

EC ($dS\ m^{-1}$)	Status	TDS ($mg\ l^{-1}$)
<0.7	Fresh water – drinking and irrigation	<500
0.7–3.0	Slightly saline – irrigation	500–2000
3.0–6.0	Medium saline – suitable for high salt tolerant crops with proper management	2000–4000
>6.0	Highly saline – not suitable for irrigation	4000
>14	Very saline – secondary drainage water	>9000
>45	Brine – sea and industrial water	>30,000

Table 1.
Irrigation water-salinity based classification.

chlorophyll content, growth yield characters of the crop. Due to high salinity of water osmotic and ion imbalance reduction in activation of enzyme responsible for the seed germination and also affects the total chlorophyll due to reduction in leaf pigments and membrane stability. Usually saline water highly influence root zone salinity [14].

5. Salinity effects

5.1 Effect on plant growth

The deficiency of ions, osmatic stress and oxidative stress to the plants. Under high saline condition for plants are not able to maintain osmotic balance leads to loss of turgidity, dehydration of cell due to higher saline environment movement of water from plant sap to soil in order to reduce the concentration gradient [15]. Due to ion toxicity and osmatic stress nutrient absorbed by mass flow greatly reduced. Due to high ion concentration affects soil osmatic potential which reduces the plant available water [16].

5.2 Effect on soil fertility

Soil nutrient availability and uptake affected by higher saline condition. Phosphorus can be affected by fixation of calcium phosphate (Ca-P) due to higher saline condition Ca ion activity is high [17]. N bioavailability and accumulation affected by soil salinity. The 25% of N requirement is fulfilled by N fixation. Soil salinity greatly reduces the N fixing rhizobacterial growth and spread due to lesser photosynthesis cause reduction in photoaccumulates [18]. Also, nitrification reduced by salinity cause increase in NH_4 content which facilitates the loss of NH_4 by volatilisation. Potassium uptake and assimilation affected by Na/Ca ratio cause reduction in K/Na ratio of the plant [19]. Salinity did not have much negative influence on micronutrient bio availability but solubility of micronutrients is little decreased in some conditions.

5.2.1 Nutrient ratios

More than concentration of the ions ratio of nutrient ions which greatly influence the bio availability and uptake of nutrient ions ratios like Na/Ca, Na/K, Ca/Mg and Cl/NO_3 [19].

6. Saline soil management

6.1 Periodical monitoring and assessment by salinity mapping

Various natural and anthropogenic factors which increase the deterioration, and severity of saline soil in order to reduce adverse effect, continuous monitoring and prediction is needed. Various quantitative and qualitative mapping done through multi-temporal and multi-spectral information from remote sensing is needed. Hyper spectral remote sensing: it is having number of narrow and continuous band provide a precise information about difference in halophytes and non halophytes and used to distinguish various salinity classes [20–22].

Number of indices used for measure salinity.

$$NDVI\left(normalised\ difference\ salinity\ index\right) = \frac{(R - NIR)}{(R + NIR)} \tag{1}$$

$$Brightness\ index = \sqrt{\mathrm{R2} + NIR.}$$

$$Salinity\ index = \sqrt{B \times R} \tag{2}$$

where NIR, near infrared reflectance; R, red band; B, blue band.

6.2 Physical management of saline soil

i. **Scrapping:** Scrapping is followed in minimal land area where removal of salt accumulated topsoil in order to minimise the root zone salinity and temporary toxic effect of salinity.

ii. **Subsoiling**: soil in deep layers having less salt content compared to above layers subsoiling breaks the top custard soil and make more permeable.

iii. **Deep ploughing**: chisel plough is needed for deep ploughing in order to increases the permeability for better leaching.

iv. **Levelling**: Levelling is to get uniform leaching entire land should be levelled avoiding unnecessary wastage of water.

v. **Sand mixing:** In heavy clay soil permeability was very less. Application and mixing of sand in soil having clay content 30–40% increases the permeability and get higher leaching efficiency.

vi. **Leaching:** Dissolve and translocate the soluble salts in downwards below 45–60 cm. Based on the water availability and soil types the leaching method differentiated into two types continuous leaching and intermittent leaching. Intermittent leaching is done by after drain of previous leaching water another application takes place which is followed in were the scarcity of good quality water.

$$Leaching\ requirement\left(LR\right) = \frac{EC_{iw}}{EC_{dW}} \times 100 \tag{3}$$

where EC_{iw} = electrical conductivity of irrigated water, EC_{dw} = electrical conductivity of drainage water.

vii. **Mulching:** Mulching with crop residues or live crops reduces the evaporation of moisture from the soil surface compared to barren soil. Its reduces the upward pull of salts from ground water table [23–25].

6.3 Chemical method

i. **Gypsum application**: Gypsum application required only the Na content of the soil increased ESP more than 15 or pH more than 8 (saline sodic soil) for replaces the Na^+ by Ca^{2+} and subsequent leaching of Na^+.

ii. **Nutrient addition**: Application of nutrient like NPK, magnesium and hormones like salicylic acid which reduces the toxicity effects of saline soil and raise optimum crop growth and yield. Nitrate which reduces the uptake of chloride, potassium reduces the uptake of Na [26]. Salicylic acid application increases Mg uptake which influences the activity of ATP leads to increase in H^+ATP-ase hydrolytic activity and imports H^+ ion in vacuole leads to increase in sodium sequestration by vacuole. K^+ foliar and soil application significantly reduces the toxic effect of saline soil by maintains the water balance and ion ratio [27].

6.4 Bio remediation of saline soil

Bioremediation is a sustainable approach in order to reduce and alleviate the toxic effect of salinity. Two different types in bio remediation are:

- Phytoremediation
- Microbial remediation

6.4.1 Phytoremediation

Phytoremediation involves using plant species to diminish the concentration of the salts or contaminants in the soil. Plant species like halophytes, hyper accumulating plants, salt tolerant and transgenic salt tolerant plants used for phyto remediation of saline soil. ex., *Tamix chinensis, Lycium chinense, Gossypium hirsutum*. There are three types of halophytes: (a) salt excluding – filters the salt by specially adoptive root system, ex. *Rhizophora muaneta*; (b) salt excreting – regulate the plant internal cell physiology, ex. *Avicennia officinalis*; salt accumulating – accumulating salts in cells and tissues of halophytes, ex. *Sonneratia apetala* [28].

6.4.2 Microbial remediation

Various salt tolerant rhizosphere microbial community (Halophills) which remediate the saline soils. Mechanism of microbial tolerant involves (i) maintains cytoplasmic ionic content equal to the medium, (ii) concentrating solutes for create osmotic balance, (iii) after cell physiology restrict or control movement of water both inside and outside of the cytoplasm [29]. *Halobacillus* sp., like *Bacillus gibsonii, Halobacterium salinarum, Staphylococcus succinus, Zhihengliuella halotolerans, Oceanobacillus oncorhynchi* are the examples of halotolerent bacteria (**Table 2**) [30].

Halophiles	Tolerate up to (M-molar salt content)
Non-halophiles	<0.2 M
Slightly halophiles	0.2–0.5 M
Moderately halophiles	0.5–2.5 M
Strongly halophiles	1.5–4.0 M
Extremely halophiles	2.5–5.2 M

Table 2.
Classification of microorganisms based on salinity tolerance [31].

6.4.3 Plant microbe interaction

Arbuscular mycorrhizal fungi (AMF) which involves in alleviate the detrimental effect of saline soil by facilitate to satisfy the plants nutrient and water requirement (**Table 3**). AMF increases the nutrient uptake of P, N, Ca, Mg and Zn, maintains the K:Na ratio, accumulate osmolytes like proline, polyamine and antioxidents, physiological changes like increase cell permeability and increase in photosynthetic efficiency finally molecular changes like maintains the activity of transporters like Pht.1 (Phosphorus transpolar) and antiportors of Na^+/H^+ (**Figure 1**) [27, 32–34].

Sensitive (0-4 d Sm–1)	Moderately sensitive (4-6 d Sm–1)	Moderately tolerant (6-8 d Sm–1)	Tolerant (8-12 d Sm–1)
Carrot	Onion	Oats	Asparagus
Cucumber	Lettuce	Cabbage	Beet root
Water melon	Potato	Safflower	Barley
Beans	Rice	Wheat	Rye
Radish	Grape	Spinach	
Pear	Soy Bean	Cotton	
Celery	Tomato		
Citrus			

Table 3.
Classification of crops based on salinity tolerance.

Figure 1.
Physical management practices of salinity management: (a) deep ploughing and (b) mulching.

7. Conclusion

Salinity is a growing threat mainly by anthropogenic activities like improper utilisation of resources. Secondary salinisation cause major factor which decline the soil fertility and leads to reduction productivity of cultivable lands. Among the various management process leaching considered more economical and efficient method. Mulching which greatly reduces further build-up of soil salinity due to reduction in evapotranspiration. Irrigation water management like monitoring salinity of irrigation water, controlled irrigation and blend with good quality water give an acceptable result on salinity management. Salinity of soil is not static, so proper monitoring and management of saline soil is required to get towards a step of global food security. A sustainable way to remediate a saline soil is still to be achievable. The knowledge regards the saline soil is needed for better work on it.

Author details

Muthuraman Yuvaraj[1*], Kasiviswanathan Subash Chandra Bose[2], Prabakaran Elavarasi[3] and Eman Tawfik[4]

1 Agricultural College and Research Institute, Tamil Nadu, India

2 Department of Soil Science and Agricultural Chemistry, Tamil Nadu Agricultural University, Coimbatore, India

3 Adhiparasakthi Agricultural College, Vellore, India

4 Genetics and Genetic Engineering, Faculty of Science, Helwan University, Egypt

*Address all correspondence to: yuvasoil@gmail.com

References

[1] Abrol IP, Yadav JSP, Massoud FI. Salt-Affected Soils and Their Management (No. 39). Rome, Italy: Food & Agriculture Organization; 1988

[2] IAB. Indian Agriculture in Brief. 27th ed. New Delhi: Agriculture Statistics Division, Ministry of Agriculture, Govt. of India; 2000

[3] Schofield RV, Kirkby MJ. Application of salinization indicators and initial development of potential global soil salinization scenario under climatic change. Global Biogeochemical Cycles. 2003;**17**(3)

[4] Wanjogu SN, Muya EM, Gicheru PT, Waruru BK. Soil degradation: Management and rehabilitation in Kenya. In: Proceedings of the FAO/ISCW Expert Consultation on Management of Degraded Soil in Southern and Eastern Africa (MADS-SEA) 2nd Networking Meeting, Pretoria, South Africa. 2001. pp. 102-113

[5] Yadav S, Irfan M, Ahmad A, Hayat S. Causes of salinity and plant manifestations to salt stress: A review. Journal of Environmental Biology. 2011;**32**(5):667

[6] Kiziloglu FM, Turan M, Sahin U, Kuslu Y, Dursun A. Effects of untreated and treated wastewater irrigation on some chemical properties of cauliflower (*Brassica olerecea* L. var. botrytis) and red cabbage (*Brassica olerecea* L. var. rubra) grown on calcareous soil in Turkey. Agricultural Water Management. 2008;**95**(6):716-724

[7] Mahmuduzzaman M, Ahmed ZU, Nuruzzaman AKM, Ahmed FRS. Causes of salinity intrusion in coastal belt of Bangladesh. International Journal of Plant Research. 2014;**4**(4A):8-13

[8] Miura K, Subhasaram T. Soil salinity after deforestation and control by reforestation in Northeast Thailand. Tropical Agricultural Research. 1991;**24**:186-196

[9] Zaman M, Shahid SA, Heng L. Irrigation systems and zones of salinity development. In: Guideline for Salinity Assessment, Mitigation and Adaptation Using Nuclear and Related Techniques. 2018. pp. 91-111

[10] Sarkar D, Haldar A. Text Book of Physical and Chemical Methods in Soil Analysis. New Delhi: New Age International Private Limited; 2011

[11] Analytical R. Conductivity Theory and Practice. France: Radiometer Analytical SAS; 2004

[12] Rhoades JD. Salinity: Electrical conductivity and total dissolved solids. In: Methods of Soil Analysis: Part 3 Chemical Methods. Vol. 5. 1996. pp. 417-435

[13] Richards LA. Diagnosis and improvement of saline and alkali soils. In: Agricultural Hand Book. Vol. 60. 1954. p. 160

[14] Zaman M, Shahid SA, Heng L. Irrigation water quality. In: Guideline for Salinity Assessment, Mitigation and Adaptation using Nuclear and Related Techniques. 2018. pp. 113-131

[15] Munns R. Genes and salt tolerance: Bringing them together. New Phytologist. 2005;**167**(3):645-663

[16] Gupta B, Huang B. Mechanism of salinity tolerance in plants: Physiological, biochemical, and molecular characterization. International Journal of Genomics. 2014

[17] Grattan SR, Grieve CM. Mineral nutrient acquisition and response by

plants grown in saline environments. In: Handbook of Plant and Crop Stress. Vol. 2. 1999. pp. 203-229

[18] Fageria NK, Gheyi HR, Moreira A. Nutrient bioavailability in salt affected soils. Journal of Plant Nutrition. 2011;**34**(7):945-962

[19] Subbarao GV, Johansen C, Jana MK, Rao JK. Effects of the sodium/calcium ratio in modifying salinity response of pigeonpea (*Cajanus cajan*). Journal of Plant Physiology. 1990;**136**(4):439-443

[20] Dutkiewicz A, Lewisa M, Ostendorf B. Mapping surface symptoms of dryland salinity with hyperspectral imagery. International Archives of the Photogrammetry, Remote Sensing and Spatial Information Sciences. 2006:34

[21] Elhag M. Evaluation of different soil salinity mapping using remote sensing techniques in arid ecosystems, Saudi Arabia. Journal of Sensors. 2016

[22] Gorji T, Tanik A, Sertel E. Soil salinity prediction, monitoring and mapping using modern technologies. Procedia Earth and Planetary Science. 2015;**15**:507-512

[23] Shahid SA, Rahman K. Soil salinity development, classification, assessment and management in irrigated agriculture. In: Handbook of Plant and Crop Stress. 3rd ed. Boca Raton: CRC Press; 2011. pp. 23-39

[24] Shahid SA, Zaman M, Heng L. Soil salinity: Historical perspectives and a world overview of the problem. In: Guideline for Salinity Assessment, Mitigation and Adaptation using Nuclear and Related Techniques. Springer; 2018. pp. 43-53. DOI: 10.1007/978-3-319-96190-3_2

[25] Siyal AA, Siyal AG, Abro Z. Salt affected soils their identification and reclamation. Paksitan Journal of Applied Science. 2001;**2**:537-540

[26] Martinez V, Cerda A. Influence of N source on rate of Cl, N, Na and K uptake by cucumber seedlings grown in saline condition. Journal of Plant Nutrition. 1989;**12**(8):971-983

[27] Golezani GK, Abriz FS. Foliar sprays of salicylic acid and jasmonic acid stimulate H^+-ATPase activity of tonoplast, nutrient uptake and salt tolerance of soybean. Ecotoxicology and Environmental Safety. 2018;**166**:18-25

[28] Wang X, Sun R, Tian Y, Guo K, Sun H, Liu X, et al. Long-term phytoremediation of coastal saline soil reveals plant species-specific patterns of microbial community recruitment. Msystems. 2020;**5**(2)

[29] Vreeland RH. Mechanisms of halotolerance in microorganisms. Critical Reviews in Microbiology. 1987;**14**(4):311-356

[30] Orhan F. Alleviation of salt stress by halotolerant and halophilic plant growth-promoting bacteria in wheat (*Triticum aestivum*). Brazilian Journal of Microbiology. 2016;**47**(3):621-627

[31] Kushner DJ. Growth and nutrition of halophilic bacteria. The Biology of Halophilic Bacteria. 1993:87-103

[32] Hammer EC, Nasr H, Pallon J, Olsson PA, Wallander H. Elemental composition of arbuscular mycorrhizal fungi at high salinity. Mycorrhiza. 2011;**21**(2):117-129

[33] Evelin H, Kapoor R, Giri B. Arbuscular mycorrhizal fungi in alleviation of salt stress: A review. Annals of Botany. 2009;**104**(7):1263-1280

[34] Romero-Munar A, Baraza E, Gulias J, Cabot C. Arbuscular mycorrhizal fungi confer salt tolerance in giant reed (*Arundo donax* L.) plants grown under low phosphorus by reducing leaf Na^+ concentration and improving phosphorus use efficiency. Frontiers in Plant Science. 2019;**10**:843

Chapter 6

Soil Erosion Influencing Factors in the Semiarid Area of Northern Shaanxi Province, China

Ning Ai, Qingke Zhu, Guangquan Liu and Tianxing Wei

Abstract

The semiarid loess area in north Shaanxi Province is one of the most serious areas of water erosion of China. Vegetation, rainfall, soil, and topography are the most dominant natural factors affecting soil erosion; and land disturbance/restoration is a significant factor influencing runoff and sediment yield in this area. According to the research, the results showed: (1) the relative impacts of the four factors on runoff were rainfall > soil > topography > vegetation, and the relative impacts of the factors on sediment yield were soil > runoff > rainfall > topography > vegetation; (2) during a period of the preliminary stage after land disturbance, topography was the most critical factor for the runoff, while sediment yield was soil. During a period of land restoration after land disturbance, runoff was primarily affected by vegetation, while sediment yield was rainfall; (3) this study showed *Hippophae rhamnoides* + *Pinus tabulaeformis* and *Hippophae rhamnoides* were the best vegetation type for reducing runoff and sediment yield, especially if the slope is less than 20°. Land disturbance is caused by human activities in semiarid regions, and in order to reduce the runoff and sediment yield quickly and effectively, artificial measures should be taken for rehabilitation of the disturbed lands.

Keywords: soil erosion, vegetation, soil and water conservation, gray relational analysis

1. Introduction

Soil erosion is one of the global environmental problems facing human survival and development. At present, the global soil erosion area is about 1.643×10^7 km^2, accounting for 10.95% of the total surface area [1, 2]. China is one of the countries with the most severe soil erosion in the world. Soil erosion has the characteristics of wide distribution area, high intensity, complex and diverse forms of erosion, and serious soil loss [3]. According to second remote sensing soil erosion survey data in China, soil erosion area was 3.56 million km^2, accounting for 37% of total land area in China; the hydraulic erosion area was 1.65 million km^2, and the wind erosion area was 1.91 million km^2; all of the erosion causes 5 billion tons of soil loss in China each year [4]. Shaanxi Province is one of the most serious areas of soil erosion in the world. Vegetation, rainfall, soil, and topography are the primary factors influencing soil erosion, although other factors may be involved. The kinetic impact of rain hitting the soil causes water erosion [5, 6], and water erosion will occur when rainfall exceeds a certain value in a single rainfall event. Many scholars have calculated a rainfall erosion standard based on research in the loess area [7–10]. Vegetation type and coverage can reduce the soil erosion index, the effectiveness of rainfall, and the kinetic energy of raindrops and runoff and lead to different soil bulk densities [11–20]. Splash from raindrops falling on the soil surface may destroy the structure of soil by causing the displacement of soil particles (splash erosion), allowing soil movement and transportation with runoff. Therefore, particle size, bulk density, initial water content, and infiltration properties of soils have important roles in water erosion and soil loss [11, 21–26]. Topography restricts the types and configuration of vegetation and affects soil moisture, runoff production, and runoff pathways, thus affecting water erosion and soil loss [22, 27–31].

Land use/cover and management are considered to be the most important factors influencing soil erosion [32–37], especially in the semiarid loess regions [11, 25, 38–41]. However, land disturbance and restoration are key factors influencing land use/cover and management [42–46]. The vegetation, root system, soil characteristics, and topography are strongly influenced by land disturbance/restoration, such as trampling and digging [46–49]. All these factors are critically important regarding runoff and sediment yield. Other human activities, such as overgrazing and deforestation, also increased the possibility of producing runoff and sediment yield [31, 50–57]. Here, we use trampling as an example to illustrate the importance of land disturbance or land mismanagement. Trampling can decrease the soil macro-porosity and the associated hydraulic conductivity, thus increasing runoff production [47, 58]. Trampling can also damage plant root system, reduce vegetation coverage, and destroy soil structure, thus rendering the soil surface more susceptible to erosion [48, 49].

In recent years, with the implementation of the Grain for Green Program and other forestry ecological engineering, a great deal of scientific attentions has been focused on the land disturbance or/and land mismanagement and their impacts on runoff production and soil erosion. Zhao et al. studied the dynamic effects of pastures and crops on runoff production and sediment yield under simulated rainfall conditions and found that vegetation restoration can reduce sediment yield more effectively at the growing stage and can reduce runoff production more effectively at the mature stage [59]. Pan et al. investigated the influence of grass and moss on runoff production and sediment yield also under simulated rainfall conditions and found that the grass and moss can efficiently reduce sediment yield and runoff production [60]. Wei et al. studied the effects of surface conditions and rainfall intensities on runoff production using micro-runoff plots and rainfall simulation and concluded that the runoff production varies drastically with different surface conditions and also with different rainfall intensities [61]. Li et al. investigated the soil detachment capacity and its relationships with sediment yield and runoff production and found that such factors as soil aggregate median diameter, organic matter, and root density can affect soil detachment capacity and thus runoff production and sediment yield [40].

However, the effects of natural rainfall events on runoff and sediment yield were found strongly different with artificial rainfall simulations [22]. Few scholars have studied the relative weights of the four primary factors that control runoff and sediment yield. Moreover, few works in the literature focused on the weights of various factors on runoff and sediment yield during the process of land disturbance/restoration under the conditions of natural rainfall. Above all, the specific objectives of this research were to: (1) Better understand the effects of the four factors—vegetation, rainfall, soil, and topography—on rainfall-runoff and sediment yield in the semiarid loess area of Shaanxi, China, and (2) examine the characteristics of annual runoff and sediment yield under different vegetation types during a period of the preliminary stage (PPS) after land disturbance and a period of land restoration after land disturbance (PLR).

2. Materials and methods

2.1 Study area

The study area is located at the field station (**Table 1** and **Figure 1**) in Dajigou catchment, a typical loess hilly area, at northwestern Loess Plateau, Shaanxi Province, China (36°33′33″–37°24′27″N, 107°38′37″–108°32′49″E; 1233–1809 MASL). The area belongs to the semiarid temperate climate zone. The mean annual precipitation is approximately 464.5 mm (1957–2013), of which approximately 61% falls in the summer from July to September. The monthly temperature ranges from −28.5°C (December 1967) to 38.3°C (July 2001), with an annual mean temperature of 7.9°C (1957–2013). The typical soils in Wuqi County are loess soils with relatively coarse particles [62]. The original vegetation almost disappeared. In recent years, to protect the environment in this region, the Chinese government implemented the Grain for Green Program to restore the ecological environment, namely, by restoring forest and grass vegetation. The major vegetation types are grasses, *Hippophae rhamnoides* (a spiny deciduous shrub), *Pinus tabulaeformis*, *Robinia pseudoacacia* (black locust), and other shrub and tree species. The shrub vegetation contains mixed deciduous broad-leaved species (i.e., *Robinia pseudoacacia* + *Hippophae rhamnoides*) and evergreen coniferous species (i.e., *Hippophae rhamnoides* + *Pinus tabulaeformis*).

2.2 Experimental design

After a complete catchment survey, together with an evaluation of topography and vegetation types, five plots (20 m × 5 m) were constructed at the study areas in July 2009. The vegetation types in the five plots are *Hippophae rhamnoides* + *Pinus tabuliformis* (I) (PRa), *Hippophae rhamnoides* + *Pinus tabuliformis* (II) (PRb), *Pinus tabuliformis* (P), *Hippophae rhamnoides* (R), and *Lespedeza davurica* + *Leymus*

Plot code	Vegetation type	Density		High	Slope gradient	Slope aspect	Elevation	Canopy density (%)	
		Pinus tabuliformis	***Hippophae rhamnoides***					**PPS**	**PLR**
RPa	*Hippophae rhamnoides* + *Pinus tabuliformis* (I)	800/hm^2	1500/hm^2	1.88 m	12°	ES37°	1396 m	48	63
RPb	*Hippophae rhamnoides* + *Pinus tabuliformis* (II)	700/hm^2	1400/hm^2	1.98 m	29°	ES35°	1380 m	32	50
P	*Pinus tabuliformis*	1200/hm^2	3000/hm^2	3.33 m	17°	WS12°	1386 m	40	53
R	*Hippophae rhamnoides*	2300/hm^2	2300/hm^2	2.62 m	17°	NE34°	1406 m	35	55
G	*Lespedeza davurica* + *Leymus secalinus*				28°	WS3°	1398 m	55	81

Table 1.
The specific conditions of five runoff plots.

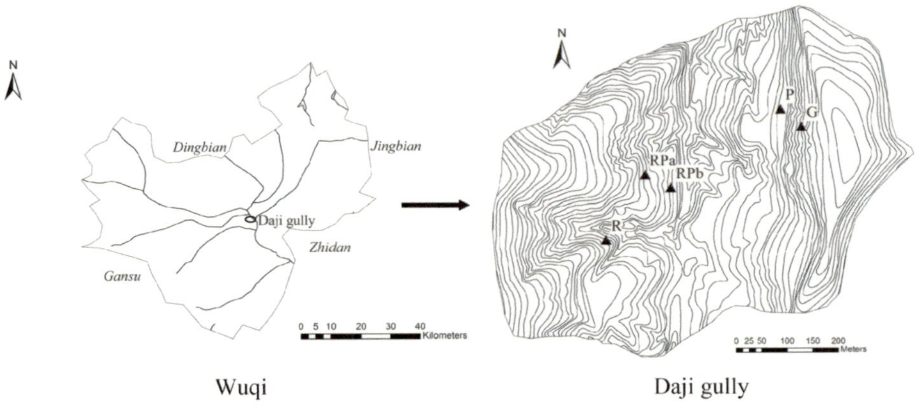

Figure 1.
The location of the study area and distribution of the five runoff plots.

secalinus (G) (**Table 1**). During the preliminary stage of establishing the plots, the vegetation, soil, and earth surface were severely degraded by the constructors through trampling and digging. With time, the vegetation and soil conditions in the plots obviously improved. The runoff and sediment yield during PPS and PLR were extremely different. This study uses the data measured in 2009 as of PPS and in 2010–2012 as of PLR for analysis. We ultimately drew conclusions regarding the order of weight for factors affecting runoff and sediment yield during the two different stages. **Figures 2** and **3** show part of the runoff plots during these two stages.

Figure 2.
The plots and meteorological station conditions during a period of the preliminary stage (PPS) after land disturbance.

Figure 3.
Contrast in plot regions and conditions during PPS and a period of land restoration (PLR) after land disturbance.

2.3 Measurement

We monitored the daily runoff and sediment yield during the main rainy season (July–September) from 2009 to 2013. **Table 1** shows the detailed parameters of these runoff plots. At the lower end of each plot, a sump was used to collect runoff and sediment yield during each rainfall-runoff event. The sump was composed of concrete with a dimension of $1 \times 1 \times 1$ m^3. The data were measured from July to September of each year. Following each rainfall event, three samples of approximately 1.65 L were removed from each sump to estimate the sediment yield.

2.4 Meteorological data

A simple meteorological field station (HOBO Weather Station, Onset Computer Co., Bourne, MA, USA), including a tilting rain gauge, was installed in the study area to record year-round meteorological data (**Figure 2**). The weather station recorded once 5 min passed. So, we calculated the I_5, I_{10}, I_{15}, and I_{30} values from the weather station record (I_5 = 5-min maximum rainfall intensity; I_{10} = 10-min maximum rainfall intensity; I_{15} = 15-min maximum rainfall intensity; I_{30} = 30-min maximum rainfall intensity).

2.5 Soil bulk density

Three soil profiles were excavated at the uphill, middle, and downhill areas which are near the runoff plots. Soil samples were collected from each profile at the depths of 0–20, 20–40, 0–60, 60–80, and 80–100 cm. Soil bulk density was tested using a ring knife (diameter, 5 cm; height, 5 cm).

2.6 Soil steady infiltration rate

An instrument for recording the process of water infiltration into the soil was employed, and the depth of infiltration was calculated by the empirical equation:

$$H = 0.19635 \times h \times \cos a \tag{1}$$

where H is the depth of infiltration, h is the change in the standing water level, and a is the slope gradient. At the beginning of the experiment, data were recorded every 10 s for 90 s; then, data were recorded every 30 s for 5 min; at the end, data were recorded once every minute. The experiment was not completed until similar measurements were observed 5–6 times [63].

2.7 Data processing

We used the Principal Coordinates Analysis (PCoA) method to convert the qualitative variables, such as vegetation and slope aspect, into quantitative variables [64]. To reduce the error of the evaluation, we used PCoA for numerical transformation and calculate the characteristic value of each item, and then we used characteristic value for analysis.

The Chinese scholar Deng Julong first proposed the gray correlation method. This method is based on developmental trends of the degree of similarity or dissimilarity between factors. By comparing a sequence of an established family of curves and a reference sequence curve, using geometric similarity, one can determine the degree of connection between the reference sequence and a comparison sequence set of data. If the shapes are similar, then a greater degree of similarity is identified. The comparison sequence and the reference sequence include both temporal series and nontemporal series. Therefore, the gray correlation method provides a quantitative measurement for the development of a system. This method is highly suitable for the analysis of a dynamic process. The specific method is shown as below [65]:

1. The parameters are standardized using:

$$x_i'(\mathrm{k}) = \frac{x_i(\mathrm{k}) - min\, x_i(\mathrm{k})}{max\, x_i(\mathrm{k}) - min\, x_i(\mathrm{k})} \tag{2}$$

where $x_i'(\mathrm{k})$ are the new values obtained when the parameters are standardized by Eq. (2) and $x_i'(\mathrm{k})$ are the original parameters.

2. Then, the correlation coefficient is calculated using:

$$\gamma\left(x_0'(\mathrm{k})\mathrm{k}, x_i'(\mathrm{k})\right) = \frac{\Delta x_{min} + \varepsilon \Delta x_{max}}{\Delta x_{0i}(k) + \varepsilon \Delta x_{max}} \tag{3}$$

$$\Delta x_{min} = \min_{\forall j \epsilon i} \min_{\forall k} \left| x_0'(\mathrm{k}) - x_j'(\mathrm{k}) \right| \tag{4}$$

$$\Delta x_{max} = \max_{\forall j \epsilon i} \max_{\forall k} \left| x_0'(\mathrm{k}) - x_j'(\mathrm{k}) \right| \tag{5}$$

$$\Delta x_{0i}(\mathrm{k}) = \left| x_0'(\mathrm{k}) - x_i'(\mathrm{k}) \right| \tag{6}$$

where $\Delta x_{0i}(\mathrm{k})$ is the absolute value of the difference between the comparison sequence and the reference sequence and ξ is the distinguishing coefficient.

		Runoff			Sediment yield		
		Correlation	**Rank**	**Proportion**	**Correlation**	**Rank**	**Proportion**
Vegetation	Vegetation types	0.5791	13	22.28%	0.5851	13	16.85%
	Vegetation coverage	0.5908	12		0.5393	14	
Rainfall	Rainfall amount	0.7685	1	27.85%	0.6922	4	20.74%
	Rainfall duration	0.704	6		0.6474	8	
	Average rainfall intensity	0.7526	2		0.8285	2	
	I_5	0.6994	7		0.643	9	
	I_{10}	0.7112	5		0.6666	7	
	I_{15}	0.7358	4		0.687	5	
	I_{30}	0.7482	3		0.6797	6	
Soil	Soil bulk density	0.6948	8	25.53%	0.8657	1	22.57%
	Soil steady infiltration rate	0.6455	10		0.6411	10	
Topography	Slope aspect	0.6655	9	24.34%	0.6315	11	18.46%
	Slope gradient	0.6124	11		0.6009	12	
Runoff					0.7134	3	21.38%

Note: I_5, 5-min maximum rainfall intensity; I_{10}, 10-min maximum rainfall intensity; I_{15}, 15-min maximum rainfall intensity; I_{30}, 30-min maximum rainfall intensity; GRG, gray relational grade

Table 2.
Gray relational grade between runoff and sediment yield and the factors influencing runoff and sediment yield.

		PPS			PLR		
		GRG	**Rank**	**Proportion**	**GRG**	**Rank**	**Proportion**
Vegetation	Vegetation type	0.6133	6	0.2476	0.6757	2	0.2594
	Vegetation cover	0.6023	7		0.6061	8	
Rainfall	Rainfall amount	0.6417	3	0.2411	0.6415	3	0.2539
	Rainfall duration	0.6303	4		0.7443	1	
	Average rainfall intensity	0.5835	10		0.5675	12	
	I_5	0.5584	12		0.6146	7	
	I_{10}	0.5662	11		0.6186	5	
	I_{15}	0.6209	5		0.6032	9	
	I_{30}	0.5406	13		0.6012	10	
Soil	Soil bulk density	0.5936	9	0.2539	0.5665	13	0.2408
	Soil steady infiltration	0.6524	2		0.6231	4	
Topography	Slope aspect	0.6681	1	0.2574	0.5995	11	0.2459
	Slope gradient	0.5955	8		0.6156	6	

Note: I_5, 5-min maximum rainfall intensity; I_{10}, 10-min maximum rainfall intensity; I_{15}, 15-min maximum rainfall intensity; I_{30}, 30-min maximum rainfall intensity; GRG, gray relational grade

Table 3.
Gray relational grade between runoff and its influential factors.

The value of ξ ranges from 0 to 1, but generally $\xi = 0.5$. $\gamma\left(x_0'(\mathrm{k})\mathrm{k}, x_i'(\mathrm{k})\right)$ is the correlation coefficient.

3. Lastly, the gray relational grade (GRG, Γ) is calculated using:

$$\Gamma = \frac{1}{n}\sum_{k=1}^{n} \gamma\left(x_0'(\mathrm{k}), x_i'(\mathrm{k})\right) \tag{7}$$

We selected runoff and sediment yield as the reference sequences; several indicators were used as comparative sequences, including vegetation type, vegetation coverage, rainfall amount, rainfall duration, average rainfall intensity (I_5, I_{10}, I_{15}, I_{30}), soil bulk density, soil steady infiltration rate, and slope aspect and gradient. Then, the gray relational grade was calculated for the reference and comparison sequences (**Tables 2** and **3**). Deng pointed out in his book that if the gray relational grade is large, then a close relationship exists between the sequence and reference parameters [65].

3. Results

3.1 The relationship between rainfall amount, rainfall-runoff, and sediment yield

Figures 4 and **5** show that if rainfall conditions are held constant, the runoff and sediment yield vary among the five runoff plots with different vegetation types. In the 16 rainfall events, relative to variations in sediment yield, variations in runoff were smaller, and the coefficient of variation was 88.26%. The coefficient of variation for sediment yield was 172.70%. Also, at the preliminary stage after runoff plots had been constructed, vegetation destroyed, and vegetation canopy lowered, the benefits of soil and water conservation were better in *Hippophae rhamnoides* + *Pinus tabuliformis* and *Hippophae rhamnoides* vegetation types. With the recovery of vegetation, the benefits of soil and water conservation increased in all of the vegetation

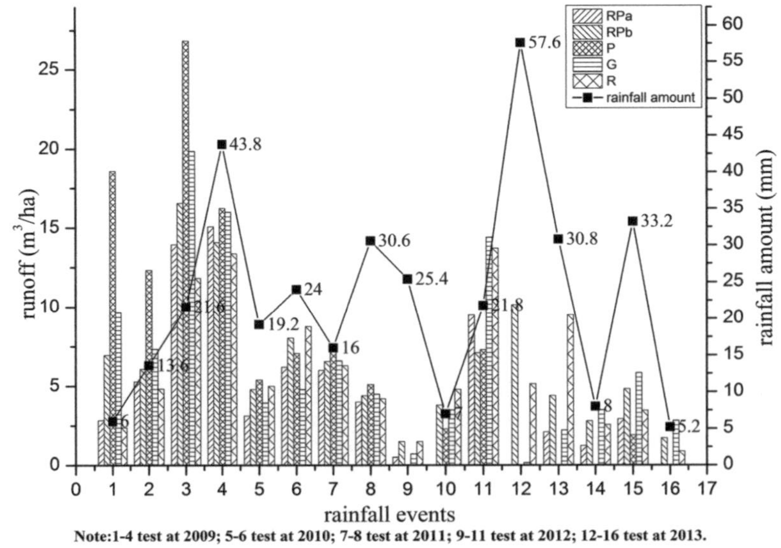

Figure 4.
Runoff trend with rainfall amounts in the study area of Wuqi County, Shaanxi Province, China.

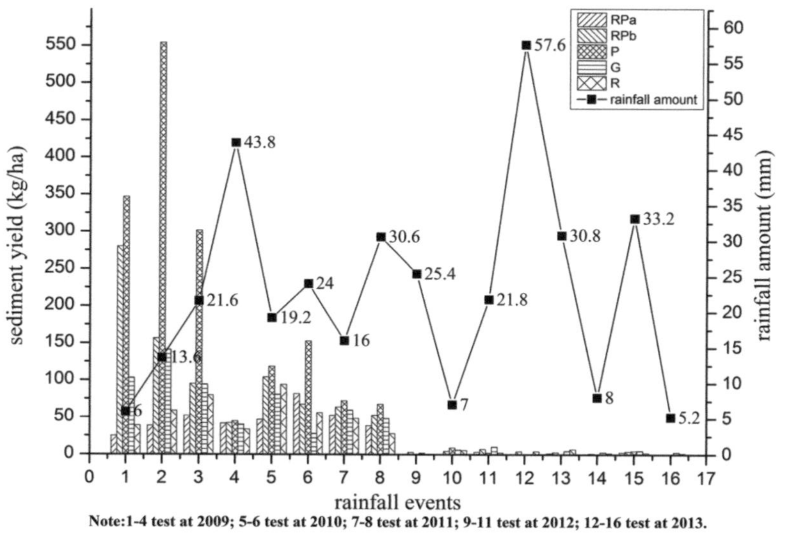

Figure 5.
Sediment yield trend with rainfall amounts in the study area of Wuqi County, Shaanxi Province, China.

types at rainfall event. However, the *Pinus tabuliformis* was more obvious, especially in low rainfall intensity and long-duration rainfall events; *Hippophae rhamnoides* + *Pinus tabuliformis* was still low at low slope gradient when contrasted with other vegetation types in runoff plots; *Hippophae rhamnoides* decreased. Comparing grassland and *Hippophae rhamnoides* + *Pinus tabuliformis* in the same slope gradient, we conclude that grass on a slope with a gradient >25° cannot take the initiative to configure arbors or shrubs and make it natural succession to grow. At same time, we suggest that some shrubs and arbors should be active configuration to enhance the effect of soil and water conservation at low slope gradient less than 25 degrees; and considering that soil and water losses in pure *Pinus tabuliformis* forest were greater in the early stage of afforestation, we especially recommend *Hippophae rhamnoides* + *Pinus tabuliformis* mixed forests.

Figures 4 and **5** show that vegetation types and rainfall amount had large effects on runoff and sediment yield; however, the change rule was not obvious. This study demonstrated that runoff and sediment yield are not solely determined by rainfall amount or by any single factor but more likely by a combination of vegetation type, vegetation coverage, rainfall amount, rainfall duration, rainfall intensity (average and for specified time periods), soil bulk density, soil steady infiltration rate, slope aspect, and slope gradient. Therefore, this research used the gray correlation method to comprehensively analyze the factors that influence runoff and sediment yield from multiple angles.

3.2 The factors affecting runoff and sediment yield based on gray relational analysis

While selecting runoff and sediment yield as a reference sequence, multiple indicators were used as comparative sequences, including vegetation type, vegetation coverage, rainfall amount, rainfall duration, average rainfall intensity, rainfall intensity for specified times (I_5, I_{10}, I_{15}, I_{30}), soil bulk density, soil steady infiltration rate, slope aspect, and slope gradient. Then the gray relational grade was calculated for the reference and comparison sequences (**Table 2**). Scientists generally agree that if the gray relational grade is large, then a close relationship exists between the sequence and reference parameters.

Several conclusions can be drawn using 5 years of data with the gray correlation method that analyzes the factors that affect runoff. First, rainfall is the most critical factor affecting runoff; it accounted for 27.86% of the total factor weight. This was followed by soil (25.53%), topography (24.34%), and vegetation (22.28%). Second, analysis of the specific factors related to rainfall found that the gray relational grade is 0.7685 for rainfall amount and that rainfall amount has the strongest influence on runoff of the top seven of 13 indicators analyzed here. Average rainfall intensity and I_{30} ranked second and third, respectively. Soil bulk density, another important factor affecting runoff, had a gray relational grade of 0.6948, which was greater than that of the soil steady infiltration rate. Slope aspect is the most important of the topographic factors affecting runoff, and its gray relational grade was 0.6655, larger than that of slope gradient. Of the vegetation factors, vegetation coverage had the largest effect on runoff and its gray relational grade, 0.5908, was larger than that of vegetation type (**Table 2**).

Several conclusions can be drawn using the gray correlation method to analyze the factors affecting sediment yield at the loess region study plots during 2009–2013. First, soil and runoff are the two most critical factors affecting sediment yield, accounting for 22.57% and 21.38% of the total proportion, while rainfall and topography accounted for 20.74% and 18.46%, respectively. Second, for the soil factors, soil bulk density had the largest effect on sediment yield and was the main factor among the 14 indicators measured here. Runoff ranked third in affecting sediment yield among the 14 indicators. Average rainfall intensity had the largest effect on sediment yield among measures of rainfall and ranked second among the 14 specific indicators. Rainfall amount also had a large effect on sediment yield, ranking fourth among the 14 indicators. The gray relational grades of other specific factors related to rainfall were also large and had dominant effects on sediment yield. The effects of vegetation type and vegetation coverage on sediment yield were small relative to other indicators; however, the gray relational grades for vegetation type and vegetation coverage were large (0.5851 and 0.5393, respectively); therefore, sediment yield and vegetation are very closely related.

3.3 Factors affecting runoff

As shown in **Table 3**, during PPS, it is clear that the slope aspect had the strongest impact on the runoff, with a GRG of 0.6681. The GRG for the soil steady infiltration rate was 0.6524, below only the slope aspect. The third factor was the rainfall amount, with a GRG of 0.6417, followed by rainfall duration, with a GRG of 0.6303. For the rainfall intensity, the GRG for I_{15} was the largest, and the relationship with runoff was similar. The influences of vegetation type and cover on the runoff were intermediate among the 13 factors and both with GRG values higher than 0.6. The smallest GRG value was I_{30} and the value was 0.5406 for the 13 factors, which is higher than 0.5. Therefore, all the 13 factors were closely related to runoff.

During PLR (**Table 3**), the rainfall duration replaced the slope aspect as the most critical factor affecting the runoff with a GRG of 0.7443. The GRG for vegetation type was 0.6757 and ranked second among the 13 factors. The rainfall amount ranked third with a GRG of 0.6415, and its influence on runoff showed no change in comparison with PLR. The GRG of the soil steady infiltration rate was ranked fourth (GRG, 0.6231), indicating a strong effect on runoff. I_{10} was the most critical factor among the rainfall intensity parameters. The influences of slope gradient and vegetation cover on runoff showed little changes. The influence of slope aspect on runoff showed a particularly notable decrease, ranking only at the 11th.

3.4 Factors affecting sediment yield

Table 4 shows that during PPS, the influence of soil bulk density was significantly higher than that of the other factors, with a GRG of 0.8113. The average rainfall intensity ranked second, with a GRG of 0.7444, followed by the soil steady infiltration rate. Among the rainfall intensity factors, I_{30} had the closest relationship with sediment yield. The GRG for vegetation type was 0.6487, indicating that this factor is more closely related to sediment yield than vegetation cover. The smallest GRG value was 0.5945, indicating that all the factors had close relationships with sediment yield.

During PLR, I_{10} replaced the soil bulk density as the most significant factor affecting the runoff, with a GRG of 0.8012. The GRG values of I_5, I_{15}, I_{30}, and average rainfall density were high among the 14 factors, indicating that rainfall intensity had the most important relationship with sediment yield during PLR. The rainfall duration ranked from the 10th to the 5th. In contrast, the soil bulk density ranked the 10th from the 1st during PPS, illustrating that the influence of the soil bulk density changed substantially during PLR. At the same time, the rank of soil steady infiltration rate was down to the eighth from the third. The rank of runoff moved up in comparison with PPS. The effects of vegetation type and vegetation cover on sediment yield were reduced according to the GRG analysis. The rank of slope aspect added, while the rank of slope gradient was down.

3.5 Runoff and sediment yield under different vegetation types

Significant differences were observed among the treatments in terms of runoff and sediment yield during PPS and PLR (**Figure 6**). The runoff and sediment yield during PPS were remarkably larger than during PLR, especially for P. During PPS, the runoff was 13.55-fold higher than that during PLR, and the sediment yield was 3.13-fold higher than that during PLR. In the analysis of grassland during PPS, the

		PPS			PLR		
		GRG	**Rank**	**Proportion**	**GRG**	**Rank**	**Proportion**
Runoff	Runoff	0.6129	13	18.62%	0.6323	11	19.32%
Vegetation	Vegetation type	0.6487	7	19.31%	0.6119	13	18.73%
	Vegetation coverage	0.6229	11		0.6141	12	
Rainfall	Rainfall amount	0.6198	12	20.22%	0.6627	9	22.54%
	Rainfall duration	0.6316	10		0.6926	5	
	Average rainfall intensity	0.7444	2		0.6758	6	
	I_5	0.6668	5		0.7953	2	
	I_{10}	0.6595	6		0.8012	1	
	I_{15}	0.6482	8		0.7779	3	
	I_{30}	0.689	4		0.7593	4	
Soil	Soil bulk density	0.8113	1	23.15%	0.6532	10	20.14%
	Soil steady infiltration rate	0.7128	3		0.6652	8	
Topography	Slope aspect	0.6362	9	18.69%	0.6658	7	19.26%
	Slope gradient	0.5945	14		0.5949	14	

Table 4.
Gray relational grade between sediment yield and its influential factors.

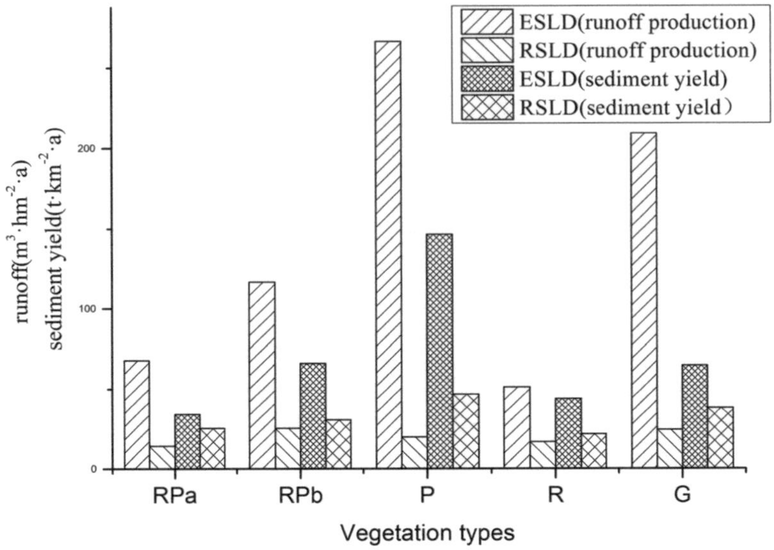

Figure 6.
Annual runoff and sediment yield during PPS and PLR for five different vegetation types.

runoff and sediment yield were 8.59-fold and 1.70-fold than those for PLR. In addition, the fold differences between PPS and PLR for the RPa, RPb, and R vegetation types in terms of runoff were 4.81, 4.60, and 3.07; the fold differences in sediment yield were 1.35, 2.17, and 2.03. This result demonstrates that vegetation recovery can effectively reduce the runoff and sediment yield.

4. Discussion

4.1 Relationship between rainfall amount, rainfall-runoff, and sediment yield

Rainfall, vegetation, soil, and topography are the main factors involved in soil erosion [66, 67]. Based on the analysis presented here (**Table 2**), rainfall amount and rainfall intensity have the greatest effect on runoff in the semiarid loess area of Shaanxi, China. This occurs because rainfall amount and intensity are closely related to the force of erosion. If the force of rainfall increases, this can potentially have a significant effect on soil loss and runoff (**Figures 4** and **5**). This conclusion, based on the data in this study, is consistent with the findings of other scholars [9, 26, 28, 68].

While the gray relational grade values of vegetation factors were large and the relationship between runoff and sediment yield was close, vegetation had the smallest influence of all the specific indicators (**Tables 2** and **3**). This may be because vegetation coverage was high during this experiment. Vegetation coverage of runoff plots was at least 32% in 2009 (**Table 1**). After 5 years of growth, the area with the least coverage had 57% vegetation cover (**Table 1**). Therefore, the effect of vegetation on runoff may be relatively low in this study area. Others have drawn the same conclusion; that is, an increase in vegetation coverage will result in a reduction in runoff so that vegetation plays a smaller role in further reducing runoff as vegetation cover increases [69–74].

In the high runoff year, after rainfall amount and intensity, topography also had a dominating influence on runoff and sediment yield (**Figures 4** and **5** and **Table 2**). This mainly occurred because different topographic conditions led to variations in soil water content in the early stage of vegetation recovery and because surface runoff differed as topography varied (**Figure 1** and **Table 1**). Low soil water content affects the infiltration capacity of soil water. If soil water content is high, soil infiltration is slow; therefore, runoff generation from excess rain leads to soil erosion [25, 26, 75–79]. The effect of rainfall intensity on runoff and sediment yield in a high runoff year was ranked from high to low, from I_5, I_{10}, I_{15}, to I_{30}. However, the ranking of the effect of rainfall intensity on runoff and sediment yield in most years was I_{30}, I_{15}, I_{10}, and I_5. Both rankings are related to the soil water content during the early stage of rainfall.

4.2 Effects of land disturbance and restoration on runoff

The weights of different factors on runoff differed significantly during PPS and PLR (**Figures 2** and **3**). **Table 3** shows that the weight significance order of the factors was topography>soil>vegetation>rainfall during PPS and the order was vegetation>rainfall>topography>soil in PLR. During PPS, the plot environments were severely degraded by trampling and digging (**Figure 2**). Slight soil disturbances do not produce serious runoff or soil erosion problems [80, 81]. In the study areas, the surface soil was destroyed, and the vegetation was heavily reduced with low vegetation coverage and canopy. The vegetation growth conditions became poor and were fragile at this time; however, stable and suitable vegetation was an effective method for reducing runoff and sediment yield [25, 82, 83].

In this study, we found that trampling and digging quantitatively decreased plant cover and vegetation, reduced soil aggregate stability, reduced soil fertility, and therefore lead to increased runoff. When the land was disturbed and the plant cover decreased, canopy interception of raindrops was low (**Figure 2** and **Table 1**). All these changes resulted in decreased mulches in the runoff plots, and thus the soil surface could not be effectively protected. This situation led to decreased rainwater infiltration and soil moisture content, and the threshold of runoff generation correspondingly decreased [39, 61, 84]. At the same time, vegetation roots were destroyed; vegetation roots can modify the structure of soil pores and can improve the soil infiltration capacity, thus reducing runoff [16, 85–88]. It has been noted that the decrease in water erosion rates with increasing root mass is exponential.

During PPS, the influence of vegetation on runoff is relatively weak, ranking the third (**Table 3**). We speculated that the protective function of vegetation on runoff was small, because during this period, all the runoff plots collected large amounts of runoff after rainfall events (**Figure 6**). Rainfall ranked fourth, for the same reason as vegetation: as long as there are rainfall-runoff events, large amounts of runoff can be produced at each runoff plot [89].

The soil surface was degraded in an irregular manner by the construction of runoff plots; therefore, the disturbances in each plot were quite different. Thus, the soil characteristics were significantly changed, especially the soil bulk density and soil steady infiltration rate. At the same time, the topography of each plot was also affected, especially the microtopography. Wilcox et al. noted that disturbances can modify surface topographical features and change the vegetation patch structure, eventually decreasing water storage within the hillslope [39]. Mohr et al. found that the impact of microtopography on surface runoff connectivity and water-repellent properties is the first-order control for hydrological and erosion processes. Therefore, during PPS, the weights of soil and topography were greater than those of vegetation and rainfall. Thus, topography and soil were major influential factors on runoff [43].

However, due to the different vegetation succession stages, the processes of runoff and soil loss are complicated and uncertain in terms of the interaction of rainfall and land use [90]. Plant growth can reduce raindrop energy and total runoff depth through canopy interception and stemflow [91, 92]. Vanacker et al. also indicated that the disturbance of vegetation cover by human activities can significantly influence erosion [46]. During PLR, the vegetation recovered, and the vegetation coverage (canopy) improved remarkably (**Table 1**). The effects of vegetation, such as the canopy interception of raindrops, the decreased velocity of raindrops, and overland flow, prevented the rainfall from directly impacting the soil surface. These effects were stronger during PLR than in PPS.

As sufficient time elapsed after the disturbance, the soil and topography became basically stable (**Figure 3**). The soils of the runoff plots consolidated and became difficult to detach by runoff. Improvements in soil characteristics such as soil porosity and organic matter increased the infiltration rate and decreased the runoff. Vegetation recovery can improve soil conditions, such as soil permeability and soil water storage after rainfall, and can control runoff loss through root-network development and litter accumulation [79, 85, 93].

Once the soil and topography were stable, the vegetation restoration and rainfall features became increasingly important for runoff. Rainfall features such as rainfall duration and rainfall intensity exhibited a strong influence on runoff generation [61, 90]. Therefore, when the soil and topography were stable and the weights of these factors on runoff were low, the weights of vegetation and rainfall on runoff increased. Vegetation was a key factor in runoff, and rainfall was the second most important factor during PLR (**Table 3**).

4.3 Effects of land disturbance and restoration on sediment yield

The weights of the studied factors on sediment yield differed significantly between PPS and PLR. The weight significance order of the factors was soil > rainfall > vegetation>topography>runoff during PPS, and the order was rainfall>-soil>runoff>topography>vegetation in PLR (**Table 4**). The sediment yield was different between PPS and PLR. During PPS, as a result of land disturbance, the sediment yield was greater than the land restoration. Through the observation on human activities, the removal of vegetation and disturbance of the soil surface result in the potential for soil structure degradation and sediment movement [94]. The sediment yield increases significantly for a short time after forest harvesting by clearcutting, and compared with good forest, the sediment yield is higher in sparse grass and bare areas which were without good cover [33]. As shown in **Table 4**, the effect of soil and rainfall on sediment yield ranked top two at both PPS and PLR. **Table 4** also shows that the weight of vegetation effect on sediment yield was the lowest at PLR, because during PLR, the vegetation cover in each plot was large (**Table 1**). Some scholars have found that a vegetation cover greater than 60% will significantly stabilize the soil surface and reduce soil erosion [49, 95]. During PPS, sediment transport capacity of the runoff was high; during PLR, sediment transport capacity of the runoff was low (**Figure 6**). Thus, the influence of runoff on sediment yield during PLR was greater than during PPS.

4.4 Influence of vegetation type on runoff and sediment yield

By estimating the annual runoff and sediment yield data for the five different vegetation types in each plot, we found that the runoff and sediment yield differed with respect to the different vegetation types. Similar results have also been observed by other scholars [25, 38, 61, 96–100].

During PPS, the order of vegetation types for producing runoff was P > G > RPb > RPa > R; the order of vegetation types for producing sediment yield was P > RPb > G > R > RPa. During PLR, the order of vegetation types for producing runoff was RPb > G > P > R > RPa; the order of vegetation types for producing sediment yield was P > G > RPb > RPa > R. Ai et al. found similar results in their investigation of nine natural rainfall events [89].

Although the runoff and sediment yield differed for the different vegetation types, the variable coefficient for PLR was lower than that for PPS (**Figure 6**). In other words, the effect of vegetation type on soil erosion is more important during land disturbance than during land restoration or a stable vegetation period. In this study, we concluded that RPa and R were better choices for land restoration or reforestation in this area, especially for slope gradients of less than 20 degrees. A study by Chen et al. indicated that pine woodland induced the largest water loss, followed by sloping cropland, alfalfa, semi-natural grassland, and shrub land, in the Loess Plateau in China [11]. Wei et al. found that shrub species were better than grass species for retaining runoff and reducing surface water loss through overland flow in a loess hilly area in China [61].

5. Conclusion

There are many factors that cause soil erosion; the runoff and sediment yield process is complicated. According to the research of soil erosion influencing factors in the semiarid area in Northern Shaanxi Province in China, the results showed:

1. The order of factors affecting runoff was rainfall > soil > topography > vegetation. Rainfall amount, average rainfall intensity, and I_{30} had the greatest effects on runoff, based on the analysis of specific indices. Rainfall indices ranked high among the 13 specific indicators. The gray relational grade values of vegetation type, which had the smallest impact on runoff among the 13 specific indicators, was 0.5791; this large value indicates a very close relationship between vegetation and runoff in the wettest year.

2. The order of factors affecting sediment yield was soil > runoff > rainfall > topography > vegetation. Soil bulk density, average rainfall intensity, and runoff had the greatest effects on sediment yield of 14 specific indicators.

3. Land disturbance and restoration significantly influence the runoff and sediment yield. The weights of influential factors (vegetation, rainfall, soil, topography) for runoff and sediment yield were also different during PPS and PLR. In this work, we determined the order of influential factor weights during PPS and PLR. This paper identified effective vegetation types for controlling runoff and reducing sediment yield. Our findings revealed that the PR and R vegetation types are better plant selections for reforestation, especially when the slope gradient is less than 20 degrees. Our research suggests that in cases of land disturbance caused by humans in semiarid regions, to quickly and effectively reduce the runoff and sediment yield, artificial measures should be taken for rehabilitation of the disturbed lands.

The results of this study will provide an important theoretical basis for the effective reduction of soil erosion during PPS and PLR, the reconstruction of low-efficiency forests, the management of spatial vegetation, and replanting of vegetation in abandoned farmlands in the semiarid loess region.

Acknowledgements

The authors would like to thank all the students and staff (Fangfang Qiang, Weijun Zhao, Huan Ma, Xingkai Zhao, Wei Qin, Zongkai Wu, Guangliang Liu, and Rui Xie) who helped in the field sampling process, especially Wu Qi Forestry Bureau.

We also thank the Engineering Research Center of Forestry Ecological Engineering, Ministry of Education, Beijing Forestry University. This study was supported by the 13th 5-year National Key Research and Development Program of China (No. 2016YFC0501602 & No. 2016YFC0501705).

Conflict of interest

The authors declare no conflict of interest.

Notes

This article is selected from *Physical Geography* and *Journal of Arid Land*, available online: http://www.tandfonline.com/ [Article DOI: http://dx.doi.org/10.1080/02723646.2015.1133189] and https://link.springe126r.com [Article DOI: https://doi.org/10.1007/s40333-016-0088-6].

Author details

Ning Ai[1,2*], Qingke Zhu[3*], Guangquan Liu[1,2*] and Tianxing Wei[3]

1 Institute of Water Resources and Hydropower Research, Beijing, China

2 College of Life Science, Yan'an University, Shaanxi Engineering and Technological Research Center for Conversation and Utilization of Regional Biological Resources, Yan'an, China

3 School of Soil and Water Conservation, Beijing Forestry University, Beijing, China

*Address all correspondence to: aining_office@126.com, xiangmub@126.com and gqliu@iwhr.com

References

[1] Ai N, Zhu Q, Liu G, Wei T. Soil Erosion Influencing Factors in the Semiarid Area of Northern Shaanxi Province, China. London: IntechOpen; 2020. DOI: 10.5772/intechopen.92979

[2] Zheng FL, Wang Z, Yang Q. The retrospection and prospect on soil erosion research in China. Chinese Journal of Nature. 2008;**30**:12-16. DOI: 10.3969/j.issn.0253-9608.2008.01.003

[3] Chen L. Soil and Water Conservation. China. 2002:4-6. DOI: 10.3969/j.issn.1000-0941.2002.07.004

[4] Li Z, Cao W, Liu B, Luo Z. Current status and developing trend of soil erosion in China. Science of Soil and Water Conservation. 2008;**6**:57-62. DOI: 10.3969/j.issn.1672-3007.2008.01.009

[5] Mieussens L, Veron F. A kinetic model for rain falling on water waves. In: EGU General Assembly Conference Abstracts. 2017. p. 17620

[6] Laburda T, Zumr D, Kavka P, Neumann M, Devaty J, Balenovic N, et al. The effects of rain drop impact on soil surface microtopography. In: 11th International Workshop on Precipitation in Urban Areas (Urban Rain 18). 2019

[7] Wang W. Study on the relations between rainfall characteristics and loss of soil in loess region. Bulletin of Soil and Water Conservation. 1983;**3**: 7-13

[8] Zhou P, Wang Z. A study on rainstorm causing soil erosion in the loess plateau. Journal of Soil and Water Conservation. 1992;**6**:1-5

[9] Wang W, Jiao J, Hao X. Statistical analysis on inter relationship among erosion sediment intensity, area, and amount in loess plateau. Journal of Soil Erosion and Soil and Water Conservation. 1998;**4**:54-60

[10] Zhang Y, Zhu Q. Statistical analysis of erosive rainfall on the loess plateau. Journal of Arid Land Resources and Environment. 2006;**20**:99-103

[11] Chen LD, Huang ZL, Gong J, Fu BJ, Huang YL. The effect of land cover/vegetation on soil water dynamic in the hilly area of the loess plateau, China. Catena. 2007;**70**:200-208. <Go to ISI>://WOS:000248605900008

[12] Sasal MC, Castiglioni MG, Wilson MG. Effect of crop sequences on soil properties and runoff on natural-rainfall erosion plots under no tillage. Soil and Tillage Research. 2010;**108**: 24-29. DOI: 10.1016/j.still.2010.03.010

[13] Zhang G-H, Liu G-B, Wang G-L. Effects of Caragana Korshinskii Kom. Cover on runoff, sediment yield and nitrogen loss. International Journal of Sediment Research. 2010;**25**:245-257. DOI: 10.1016/S1001-6279(10)60042-X

[14] Xiao P, Yao W, Römkens MJM. Effects of grass and shrub cover on the critical unit stream power in overland flow. International Journal of Sediment Research. 2011;**26**:387-394. DOI: 10.1016/S1001-6279(11)60102-9

[15] Zokaib S, Naser G. Impacts of land uses on runoff and soil erosion: A case study in Hilkot watershed Pakistan. International Journal of Sediment Research. 2011;**26**:343-352. DOI: 10.1016/S1001-6279(11)60098-X

[16] Wang C, Zhao C, Xu Z, Wang Y, Peng H. Effect of vegetation on soil water retention and storage in a semi-arid alpine forest catchment. Journal of Arid Land. 2013;**5**:207-219. DOI: 10.1007/s40333-013-0151-5

[17] Zhang B, Zhang G, Yang H, Zhu P. Temporal variation in soil erosion resistance of steep slopes restored with different vegetation communities on the

Chinese loess plateau. Catena. 2019;**182**: 104170

[18] Chen J, Xiao H, Li Z, Liu C, Wang D, Wang L, et al. Threshold effects of vegetation coverage on soil erosion control in small watersheds of the red soil hilly region in China. Ecological Engineering. 2019;**132**:109-114

[19] Wu Z, Wang M, Zhang H, Du Z. Vegetation and soil wind erosion dynamics of sandstorm control programs in the agro-pastoral transitional zone of northern China. Frontiers of Earth Science. 2019;**13**: 430-443

[20] Zhou J, Fu B, Gao G, Lü Y, Liu Y, Lü N, et al. Effects of precipitation and restoration vegetation on soil erosion in a semi-arid environment in the loess plateau, China. Catena. 2016;**137**:1-11

[21] Martinez-Mena M, Rogel JA, Albaladejo J, Castillo VM. Influence of vegetal cover on sediment particle size distribution in natural rainfall conditions in a semiarid environment. Catena. 2000;**38**:175-190. DOI: 10.1016/S0341-8162(99)00073-9

[22] Mathys N, Klotz S, Esteves M, Descroix L, Lapetite JM. Runoff and erosion in the black marls of the French Alps: Observations and measurements at the plot scale. Catena. 2005;**63**: 261-281. <Go to ISI>://WOS: 000233398300008. Available from: http://www.sciencedirect.com/science/article/pii/S0341816205000949

[23] Fernández C, Vega JA, Fonturbel T, Jiménez E, Pérez JR. Immediate effects of prescribed burning, chopping and clearing on runoff, infiltration and erosion in a shrubland area in Galicia (NW Spain). Land Degradation & Development. 2008;**19**:502-515. DOI: 10.1002/ldr.855

[24] Ekwue EI, Harrilal A. Effect of soil type, peat, slope, compaction effort and their interactions on infiltration, runoff and raindrop erosion of some Trinidadian soils. Biosystems Engineering. 2010;**105**:112-118. DOI: 10.1016/j.biosystemseng.2009.10.001

[25] Mohammad AG, Adam MA. The impact of vegetative cover type on runoff and soil erosion under different land uses. Catena. 2010;**81**:97-103. DOI: 10.1016/j.catena.2010.01.008

[26] Defersha MB, Melesse AM. Effect of rainfall intensity, slope and antecedent moisture content on sediment concentration and sediment enrichment ratio. Catena. 2012;**90**:47-52. DOI: 10.1016/j.catena.2011.11.002

[27] Kinnell P. The effect of slope length on sediment concentrations associated with side-slope erosion. Soil Science Society of America Journal. 2000;**64**: 1004-1008

[28] Chaplot VAM, Le Bissonnais Y. Runoff features for interrill erosion at different rainfall intensities, slope lengths, and gradients in an agricultural loessial hillslope. Soil Science Society of America Journal. 2003;**67**:844-851. DOI: 10.2136/sssaj2003.8440

[29] DiBiase RA, Whipple KX. The influence of erosion thresholds and runoff variability on the relationships among topography, climate, and erosion rate. Journal of Geophysical Research: Earth Surface. 2011;**116**:F04036. DOI: 10.1029/2011JF002095

[30] Nadal-Romero E, Lasanta T, García-Ruiz JM. Runoff and sediment yield from land under various uses in a Mediterranean mountain area: Long-term results from an experimental station. Earth Surface Processes and Landforms. 2013;**38**:346-355

[31] Taye G, Poesen J, Van Wesemael B, Vanmaercke M, Teka D, Deckers J, et al.

Effects of land use, slope gradient, and soil and water conservation structures on runoff and soil loss in semi-arid northern Ethiopia. Physical Geography. 2013;**34**:236-259. <Go to ISI>://WOS: 000328506300005

[32] Chen L, Wang J, Fu B, Qiu Y. Land-use change in a small catchment of northern loess plateau, China. Agriculture, Ecosystems & Environment. 2001;**86**:163-172. Available from: http://ac.els-cdn.com/ S0167880900002711/1-s2.0- S0167880900002711-main.pdf?_tid= 615f3a1a-d3c7-11e4-94f9-00000aacb 361&acdnat=1427381581_77ea93fcb6d 17dec4c6bcfa67d6c9836

[33] Porto P, Walling DE, Callegari G. Investigating the effects of afforestation on soil erosion and sediment mobilisation in two small catchments in southern Italy. Catena. 2009;**79**:181-188. <Go to ISI>://WOS: 000272019300002

[34] Tiwari KR, Sitaula BK, Bajracharya RM, Borresen T. Runoff and soil loss responses to rainfall, land use, terracing and management practices in the Middle Mountains of Nepal. Acta Agriculturae Scandinavica Section B: Soil and Plant Science. 2009;**59**:197-207. <Go to ISI>://WOS:000265938000001

[35] Garcia-Ruiz JM. The effects of land uses on soil erosion in Spain: A review. Catena. 2010;**81**:1-11. <Go to ISI>:// WOS:000276064100001

[36] Nunes AN, De Almeida AC, Coelho COA. Impacts of land use and cover type on runoff and soil erosion in a marginal area of Portugal. Applied Geography. 2011;**31**:687-699

[37] Peng T, Wang SJ. Effects of land use, land cover and rainfall regimes on the surface runoff and soil loss on karst slopes in Southwest China. Catena. 2012;**90**:53-62. <Go to ISI>://WOS: 000300070300006

[38] Fu B, Chen L, Ma K, Zhou H, Wang J. The relationships between land use and soil conditions in the hilly area of the loess plateau in northern Shaanxi, China. Catena. 2000;**39**:69-78. Available from: http://www.scienced irect.com/science/article/pii/ S0341816299000843

[39] Wilcox BP, Breshears DD, Allen CD. Ecohydrology of a resource-conserving semiarid woodland: Effects of scale and disturbance. Ecological Monographs. 2003;**73**:223-239. <Go to ISI>://WOS: 000183500200004

[40] Li ZW, Zhang GH, Geng R, Wang H, Zhang XC. Land use impacts on soil detachment capacity by overland flow in the loess plateau, China. Catena. 2015;**124**:9-17. DOI: 10.1016/j. catena.2014.08.019

[41] Zhu JL, Gao P, Geissen V, Maroulis J, Ritsema CJ, Mu XM, et al. Impacts of rainfall and land use on sediment regime in a semi-arid region: Case study of the Wuqi catchment in the upper Beiluo River basin, China. Arid Land Research and Management. 2015;**29**:1-16. <Go to ISI>://WOS: 000343828400001

[42] Ungar ED, Stavi I, Lavee H, Sarah P. Effects of livestock traffic on rock fragment movement on hillsides in a semiarid patchy rangeland. Land Degradation & Development. 2010;**21**: 92-99. <Go to ISI>://WOS: 000276670700004

[43] Mohr CH, Coppus R, Iroume A, Huber A, Bronstert A. Runoff generation and soil erosion processes after clear cutting. Journal of Geophysical Research-Earth Surface. 2013;**118**:814-831. <Go to ISI>://WOS: 000324993900030

[44] Brunbjerg AK, Svenning JC, Ejrnaes R. Experimental evidence for disturbance as key to the conservation of dune grassland. Biological

Conservation. 2014;**174**:101-110. DOI: 10.1016/j.biocon.2014.04.002

[45] Malowerschnig B, Sass O. Long-term vegetation development on a wildfire slope in Innerzwain (Styria, Austria). Journal of Forestry Research. 2014;**25**:103-111. <Go to ISI>://WOS: 000330202600012. http://download.springer.com/static/pdf/794/art%253A10.1007%252Fs11676-014-0435-4.pdf?auth66=1416711293_3c3f250711475930aaf0426dbc73707a&ext=.pdf; http://link.springer.com/article/10.1007%2Fs11676-014-0435-4

[46] Vanacker V, Bellin N, Molina A, Kubik PW. Erosion regulation as a function of human disturbances to vegetation cover: A conceptual model. Landscape Ecology. 2014;**29**:293-309. <Go to ISI>://WOS: 000331935100009

[47] Herrick JE, Van Zee JW, Belnap J, Johansen JR, Remmenga M. Fine gravel controls hydrologic and erodibility responses to trampling disturbance for coarse-textured soils with weak cyanobacterial crusts. Catena. 2010;**83**: 119-126. DOI: 10.1016/j.catena.2010.08.007

[48] Dunne T, Western D, Dietrich WE. Effects of cattle trampling on vegetation, infiltration, and erosion in a tropical rangeland. Journal of Arid Environments. 2011;**75**:58-69. DOI: 10.1016/j.jaridenv.2010.09.001

[49] Pohl M, Graf F, Buttler A, Rixen C. The relationship between plant species richness and soil aggregate stability can depend on disturbance. Plant and Soil. 2012;**355**:87-102. DOI: 10.1007/s11104-011-1083-5

[50] Ludwig JA, Wilcox BP, Breshears DD, Tongway DJ, Imeson AC. Vegetation patches and runoff-erosion as interacting ecohydrological processes in semiarid landscapes. Ecology. 2005;**86**:288-297. <Go to ISI>://WOS: 000227634400003

[51] Snyman HA. Rangeland degradation in a semi-arid South Africa—I: Influence on seasonal root distribution, root/shoot ratios and water-use efficiency. Journal of Arid Environments. 2005;**60**: 457-481. <Go to ISI>://WOS: 000226455000007. Available from: http://www.sciencedirect.com/science/article/pii/S0140196304001260

[52] Snyman HA, Du Preez CC. Rangeland degradation in a semi-arid South Africa—II: Influence on soil quality. Journal of Arid Environments. 2005;**60**:483-507

[53] Lin Y, Han GD, Zhao ML, Chang SX. Spatial vegetation patterns as early signs of desertification: A case study of a desert steppe in Inner Mongolia, China. Landscape Ecology. 2010;**25**:1519-1527. <Go to ISI>://WOS: 000283371000005. Available from: http://download.springer.com/static/pdf/470/art%253A10.1007%252Fs10980-010-9520-z.pdf?auth66=1427599333_53e15338c1367d98edc8cfe209611ec6&ext=.pdf

[54] Zhao G, Mu X, Wen Z, Wang F, Gao P. Soil erosion, conservation, and eco-environment changes in the loess plateau of China. Land Degradation & Development. 2013;**24**:499-510. <Go to ISI>://WOS:000325458600009

[55] Ma L, Teng YG, Shangguan ZP. Ecohydrological responses to secondary natural *Populus davidiana* and plantation *Pinus tabulaeformis* woodlands on the loess plateau of China. Ecohydrology. 2014;**7**:612-621. <Go to ISI>://WOS:000334671600034. Available from: http://onlinelibrary.wiley.com/doi/10.1002/eco.1382/abstract

[56] Kairis O, Karavitis C, Salvati L, Kounalaki A, Kosmas K. Exploring the impact of overgrazing on soil erosion and land degradation in a dry Mediterranean agro-forest landscape (Crete, Greece). Arid Land Research and Management. 2015;**29**:360-374. <Go to ISI>://WOS:000347533900006

[57] Rather MA, Kumar JS, Farooq M, Rashid H. Assessing the influence of watershed characteristics on soil erosion susceptibility of Jhelum basin in Kashmir Himalayas. Arabian Journal of Geosciences. 2017;**10**:59

[58] McDowell RW, Drewry JJ, Paton RJ, Carey PL, Monaghan RM, Condron LM. Influence of soil treading on sediment and phosphorus losses in overland flow. Australian Journal of Soil Research. 2003;**41**:949-961. <Go to ISI>://WOS: 000185266100010

[59] Zhao XN, Huang J, Wu PT, Gao XD. The dynamic effects of pastures and crop on runoff and sediments reduction at loess slopes under simulated rainfall conditions. Catena. 2014;**119**:1-7. <Go to ISI>://WOS:000336705500001

[60] Pan CZ, Shangguan ZP, Lei TW. Influences of grass and moss on runoff and sediment yield on sloped loess surfaces under simulated rainfall. Hydrological Processes. 2006;**20**: 3815-3824. <Go to ISI>://WOS: 000242663600002

[61] Wei W, Jia FY, Yang L, Chen LD, Zhang HD, Yu Y. Effects of surficial condition and rainfall intensity on runoff in a loess hilly area, China. Journal of Hydrology. 2014;**513**:115-126. <Go to ISI>://WOS:000337258900012

[62] Wuqi Chronicle. Xi'an, China: Sanqin Press; 1991

[63] Zhao J, Wei T, Chen Z, Zhu W-d, Zhou Y. Soil and water conservation benefit analysis for different vegetation in young forests converted by farmland of Wuqi County in Shaanxi Province. Journal of Soil and Water Conservation. 2010;**24**(3):31-34

[64] Zhang J. Quantitative Ecology. Science Press; 2004

[65] Deng J-L. Theory of Grey System. Huazhong University of Science and Technology Press; 1990

[66] Ai N, Wei T, Zhu Q, Qiang F, Ma H, Qin W. Impacts of land disturbance and restoration on runoff production and sediment yield in the Chinese loess plateau. Journal of Arid Land. 2017;**9**: 76-86

[67] Gu C, Mu X, Gao P, Zhao G, Sun W, Tan X. Distinguishing the effects of vegetation restoration on runoff and sediment generation on simulated rainfall on the hillslopes of the loess plateau of China. Plant and Soil. 2020; **447**:393-412

[68] Zhang W, Xie Y, Liu B. Research evolution of rainfall erosivity. Journal of Soil and Water Conservation. 2002;**16**: 43-46

[69] Zhang X, Yu X, Wu S, Wei T. Effects of forest vegetation on runoff and sediment production on sloping lands of loess area. The Journal of Applied Ecology [Ying yong sheng tai xue bao]. 2005;**16**:1613-1617

[70] Yu X, Zhang X, Wu S, Wei T, Zhang X. The effect of vegetation and precipitation upon runoff and sediment production in sloping lands of loess area. Journal of Mountain Science. 2006;**24**(1): 19-26

[71] Foltz R, Copeland N, Elliot W. Reopening abandoned forest roads in northern Idaho, USA: Quantification of runoff, sediment concentration, infiltration, and interrill erosion parameters. Journal of Environmental Management. 2009;**90**:2542-2550

[72] Zhu B, Li Z, Li P, You Z. Effect of grass coverage on sediment yield of rain on slope. Acta Pedologica Sinica. 2010; **47**:401-407

[73] Yuehong C, Feixin W, Guangquan L, Xinxiao Y, Guodong J, Ping G. Modified vegetation-erosion dynamics model and its application in typical watersheds in the loess plateau. International Journal of Sediment Research. 2011;**26**:78-86

[74] Gao Z, Fu Y, Li Y, Liu J, Chen N, Zhang X. Trends of streamflow, sediment load and their dynamic relation for the catchments in the middle reaches of the Yellow River over the past five decades. Hydrology and Earth System Sciences. 2012;**16**:3219

[75] Léonard J, Ancelin O, Ludwig B, Richard G. Analysis of the dynamics of soil infiltrability of agricultural soils from continuous rainfall-runoff measurements on small plots. Journal of Hydrology. 2006;**326**:122-134

[76] Liu J-T, Feng D-Z, Chen X, Wu J-C, Gu W-M. Distribution characteristics of hillslope curvature and its effects on hydrological processes: A real-world test. Advances in Water Science. 2011; **22**:1-6

[77] Huang J, Zhao X-N, Wu P-T. Factors analysis of slope runoff and sediment based on path analysis and grey relational analysis. Journal of Sichuan University: Engineering Science Edition. 2012;**44**:64-70

[78] Yao X, Fu B, Lü Y. Spatial patterns of soil moisture at transect scale in the loess plateau of China. Acta Ecologica Sinica. 2012;**32**:4961-4968

[79] Huang J, Wu PT, Zhao XN. Effects of rainfall intensity, underlying surface and slope gradient on soil infiltration under simulated rainfall experiments. Catena. 2013;**104**:93-102. <Go to ISI>://WOS:000315556600009

[80] Cram DS, Baker TT, Fernald AG, Madrid A, Rummer B. Mechanical thinning impacts on runoff, infiltration, and sediment yield following fuel reduction treatments in a southwestern dry mixed conifer forest. Journal of Soil and Water Conservation. 2007;**62**: 359-366. <Go to ISI>://WOS: 000249806500022

[81] Wang XY, Gao HW, Tullberg JN, Li HW, Kuhn N, Mchugh AD, et al. Traffic and tillage effects on runoff and soil loss on the loess plateau of northern China. Australian Journal of Soil Research. 2008;**46**:667-675. <Go to ISI>://WOS:000261286500005

[82] Dunjo G, Pardini G, Gispert M. The role of land use-land cover on runoff generation and sediment yield at a microplot scale, in a small Mediterranean catchment. Journal of Arid Environments. 2004;**57**:239-256. DOI: 10.1016/S0140-1963(03)00097-1

[83] Zhang B, Yang YS, Zepp H. Effect of vegetation restoration on soil and water erosion and nutrient losses of a severely eroded clayey Plinthudult in southeastern China. Catena. 2004;**57**: 77-90. DOI: 10.1016/j.catena.2003.07.001

[84] Zhang GH, Tang MK, Zhang XC. Temporal variation in soil detachment under different land uses in the loess plateau of China. Earth Surface Processes and Landforms. 2009;**34**: 1302-1309. <Go to ISI>://WOS: 000268117800010

[85] Gyssels G, Poesen J, Bochet E, Li Y. Impact of plant roots on the resistance of soils to erosion by water: A review. Progress in Physical Geography. 2005; **29**:189-217. <Go to ISI>://WOS: 000229476500004

[86] Wang B, Zhang GH, Shi YY, Zhang XC, Ren ZP, Zhu LJ. Effect of natural restoration time of abandoned farmland on soil detachment by overland flow in the loess plateau of China. Earth Surface Processes and Landforms. 2013;**38**:1725-1734. <Go to ISI>://WOS:000328330400009

[87] Wang B, Zhang GH, Shi YY, Zhang XC. Soil detachment by overland flow under different vegetation restoration models in the loess plateau of China. Catena. 2014;**116**:51-59. <Go to ISI>://WOS:000332356200005

[88] Wang B, Zhang G-H, Zhang X, Li Z-W, Su Z-L, Yi T, et al. Effects of near soil surface characteristics on soil detachment by overland flow in a natural succession grassland. Soil Science Society of America Journal. 2014;**78**:589-597

[89] Ai N, Wei TX, Zhu QK. The effect of rainfall for runoff-erosion-sediment yield under the different vegetation types in loess plateau of northern Shaanxi Province. Journal of Soil and Water Conversion. 2013;**27**:26-30 (in Chinese)

[90] Wei W, Chen LD, Fu BJ, Huang ZL, Wu DP, Gui LD. The effect of land uses and rainfall regimes on runoff and soil erosion in the semi-arid loess hilly area, China. Journal of Hydrology. 2007;**335**: 247-258. <Go to ISI>://WOS: 000245300300003. Available from: http://www.sciencedirect.com/science/article/pii/S002216940600624X

[91] Navar J. Stemflow variation in Mexico's northeastern forest communities: Its contribution to soil moisture content and aquifer recharge. Journal of Hydrology. 2011;**408**:35-42. <Go to ISI>://WOS:000295607000003

[92] Yang FS, Cao MM, Li HE, Wang XH, Bi CF. Simulation of sediment retention effects of the single seabuckthorn flexible dam in the Pisha sandstone area. Ecological Engineering. 2013;**52**:228-237. <Go to ISI>://WOS: 000316827200029

[93] Descheemaeker K, Muys B, Nyssen J, Poesen J, Raes D, Haile M, et al. Litter production and organic matter accumulation in exclosures of the Tigray highlands, Ethiopia. Forest Ecology and Management. 2006;**233**: 21-35. <Go to ISI>://WOS: 000240409800003. Available from: http://www.sciencedirect.com/science/article/pii/S0378112706003859

[94] Vacher CA, Loch RJ, Raine SR. Effect of polyacrylamide additions on infiltration and erosion of disturbed lands. Australian Journal of Soil Research. 2003;**41**:1509-1520. <Go to ISI>://WOS:000187341700007

[95] Martin C, Pohl M, Alewell C, Korner C, Rixen C. Interrill erosion at disturbed alpine sites: Effects of plant functional diversity and vegetation cover. Basic and Applied Ecology. 2010; **11**:619-626. <Go to ISI>://WOS: 000286795300008

[96] Vacca A, Loddo S, Ollesch G, Puddu R, Serra G, Tomasi D, et al. Measurement of runoff and soil erosion in three areas under different land use in Sardinia (Italy). Catena. 2000;**40**:69-92. Available from: http://www.sciencedirect.com/science/article/pii/S0341816200000886

[97] Meng QH, Fu BJ, Tang XP, Ren HC. Effects of land use on phosphorus loss in the hilly area of the loess plateau, China. Environmental Monitoring and Assessment. 2008;**139**:195-204. <Go to ISI>://WOS:000253754000016. http://download.springer.com/static/pdf/424/art%253A10.1007%252Fs10661-007-9826-8.pdf?auth66=1427599321_fc1b632056ff9fa099183d6eb4e08060&ext=.pdf

[98] Xu XL, Ma KM, Fu BJ, Song CJ, Liu W. Influence of three plant species with different morphologies on water runoff and soil loss in a dry-warm river valley, SW China. Forest Ecology and Management. 2008;**256**:656-663. <Go to ISI>://WOS:000258361200019. Available from: http://www.sciencedirect.com/science/article/pii/S0378112708004210

[99] Xu XL, Ma KM, Fu BJ, Liu W, Song CJ. Soil and water erosion under different plant species in a semiarid river valley, SW China: The effects of plant morphology. Ecological Research. 2009;**24**:37-46. <Go to ISI>://WOS: 000261956100005. Available from: http://download.springer.com/static/

pdf/975/art%253A10.1007%252Fs
11284-008-0479-z.pdf?auth66=
1427599338_a0bae994748e7a43f7c4bd
5215a7a0f1&ext=.pdf

[100] Wang L, Wei SP, Horton R, Shao MA. Effects of vegetation and slope aspect on water budget in the hill and gully region of the loess plateau of China. Catena. 2011;**87**:90-100. <Go to ISI>://WOS:000293105300010. Available from: http://ac.els-cdn.com/
S0341816211001007/1-s2.0-
S0341816211001007-main.pdf?_tid=
0601e30c-d5c0-11e4-8ef9-00000aab0f
27&acdnat=1427598324_330f8cbaef
708153f02f74dffb9418a0